A VOICE FOR SCIENCE IN THE SOUTH

A VOICE FOR SCIENCE IN THE SOUTH

Edited by

Daniel Schaffer

The World Academy of Sciences
for the advancement of science in developing countries
Trieste, Italy

NEW JERSEY • LONDON • SINGAPORE • BEIJING • SHANGHAI • HONG KONG • TAIPEI • CHENNAI • TOKYO

Published by

World Scientific Publishing Co. Pte. Ltd.
5 Toh Tuck Link, Singapore 596224
USA office: 27 Warren Street, Suite 401-402, Hackensack, NJ 07601
UK office: 57 Shelton Street, Covent Garden, London WC2H 9HE

British Library Cataloguing-in-Publication Data
A catalogue record for this book is available from the British Library.

TWAS
A Voice for Science in the South

ISBN 978-981-4740-42-5
ISBN 978-981-4740-27-2 (pbk)

Printed in Singapore

Preface

A Voice for Science in the South *has been assembled to commemorate a pair of 30th anniversaries: the founding of the Third World Academy of Sciences (TWAS) in 1983 and the Academy's first General Meeting in 1985.*

The following essays have been written by scientists who have direct ties to the Academy. Some have long-standing connections that date back to TWAS's earliest days. Others were intensely involved with the Academy over a period of time and then moved on to other things. Still others have only recently arrived in positions of leadership.

What these scientists have in common is a deep appreciation for the Academy's mission and work. They view TWAS as an invaluable player in efforts to build scientific capacity and excellence in the developing world – and, more generally, as an organization that has helped to foster positive change for both science and society on a global scale.

The contributors pay homage to the enormous influence of Abdus Salam, TWAS's founding president and one of the most eminent scientists of the 20th century. Yet they are also willing to examine the Academy's present-day challenges and to consider the steps that they think should be taken to keep TWAS strong and sustainable in the future. Specifically, they explore the changed context in which TWAS operates, marked by vast increases in scientific capacity in the South and a widening gap between scientifically proficient countries and scientifically lagging countries. They also call on the Academy to expand and amplify its efforts to support both women scientists and young scientists as critical aspects of its efforts to build scientific capacity in the developing world in the years ahead.

Today, TWAS is The World Academy of Sciences for the advancement of science in the developing world. But the core principles and strategies that have defined the organization since its inception remain unchanged. The dream embedded in its founding, the early struggles to become relevant, the great successes and today's challenges offer deep insights into the evolution of science in the developing world over the past three decades.

This is the spirit – part celebratory and part cerebral – that infuses the essays in this collection. In telling the TWAS story from their personal perspectives, the contributors hope that they will not only shed light on the past, but also raise important points of discussion that will help to guide the Academy and to advance science in the South in the future.

Daniel Schaffer
September 2015

Contents

Foreword

TWAS, past and future

Bai Chunli

Bai Chunli has achieved extraordinary distinction in his scientific career – as a researcher and teacher, as a leader of major science organizations and as an advocate of international science cooperation. He has served since 2011 as president of the Chinese Academy of Sciences (CAS) and since 2013 as president of The World Academy of Sciences (TWAS).

Bai was born in 1953 in Dandong, Liaoning Province, located in the northeastern corner of China. At an early age, he was recognized as an outstanding student and promising young leader. After high school, he worked in a production and construction corps in Inner Mongolia. He then graduated from Department of Chemistry at Peking University and earned a master's and a doctorate degree at the CAS Institute of Chemistry. From 1985 to 1987, he conducted research in the field of physical chemistry at the California Institute of Technology. He then returned to the CAS Institute of Chemistry, and from 1991-1992 he served as a visiting professor at the Institute for Materials Research (IMR) at Tohoku University in Japan.

In the mid-1980s, Bai's research interest gravitated to nanotechnology and nano-microscopy. He developed China's earliest tools for manipulating single atoms and molecules, including the nation's first atomic force microscope, scanning tunneling microscope (STM), low-temperature STM, ultra-high vacuum-STM, and ballistic electron emission microscopy. His work further focused on molecular nanostructure and self-assembly, molecular nanodevices and novel nanomaterials. At the same time, he established methodologies for the study of molecular assembly on solid substrate surfaces in mild chemical and environmental conditions.

Together, these technologies have had enormously valuable application in materials science and other fields.

Bai is recognized in China and beyond for these landmark accomplishments. His innovations have led to a number of patents. In 2002, he was named one of China's top 10 pioneers of science and technology. He was elected to TWAS in 1997. He is also a member of a number of national science academies and societies, including the United States National Academy of Sciences in 2006 and the Royal Society in the United Kingdom in 2014.

The Chinese Academy of Sciences is seen as the engine of China's research and technology enterprise. As CAS president, Bai oversees more than 100 research institutes, two universities, and several science-related business and publishing enterprises.

He is the first Chinese scientist to serve as president of TWAS. Under his direction, the two academies have cooperated on significant new initiatives for the promotion of science in the developing world, including the launch of the CAS-TWAS President's PhD Fellowships and the creation of five CAS-TWAS Centres of Excellence based in Beijing.

***

In the 1970s and early 1980s, the world was starkly divided between nations that had advanced science and those that did not. Wealthy nations in Western Europe and North America, along with the Soviet Union and Japan, led the world in fields from chemistry and physics to medicine and space exploration. These efforts helped to drive historic levels of economic development and wealth-creation in select countries in the North. Third World nations, as they were known then, were seen as places that could achieve only limited competence in science and engineering.

Those years were dominated by the geopolitical dynamics and conventions that had taken shape after World War II. Yet looking back, we can see that in many poor nations, science even then was in ferment.

In the 1960s, the agricultural advances of the "green revolution" were dramatically improving food production in nations as diverse as India, the Philippines and Mexico. India had put a satellite into space in the 1960s, and in 1974, it had developed a nuclear bomb. In that same year, Egyptian leaders began to consider a plan to revive the *Bibliotheca Alexandrina*, which in antiquity had been a renowned centre of science and scholarship. The Kuwait Foundation for the Advancement of Sciences was founded in 1976, and in 1979, the Kuwait Institute for Scientific Research started its first five-year plan. Brazil, concerned about the risks of dependence on foreign oil, in 1975 initiated a programme that would eventually make it a world leader in the production of bioethanol from sugar cane.

In China, too, this period was the beginning of a transformation. In the mid-1960s, Premier Zhou Enlai had advocated the "Four Modernizations" to upgrade the nation's strength in agriculture, industry, defence, and science and technology. In the 1970s, leaders from China and the United States agreed to scientific exchanges as part of a larger opening to the world, and in 1978, new leader Deng Xiaoping began to implement the Four Modernizations, reflecting a profound commitment to development through science and engineering.

Certainly, however, we must recognize how Abdus Salam, the Pakistani physicist, helped to spread these ideas and galvanize these developments into a global movement. With Italian counterpart Paolo Budinich, Salam led the campaign to establish the International Centre for Theoretical Physics (ICTP), which opened in 1964 in Trieste, Italy. And in 1983, just four years after Salam won the Nobel Prize, they worked with a cadre of elite scientists from the South to found the Third World Academy of Sciences (TWAS), an advocate for science and a champion of scientific excellence that would have the unique perspective of the South.

New ambitions, new expectations

Salam at that point had been director of ICTP for nearly two decades. He had a well-honed feel for the needs of that era – the needs of nations, and the needs of scientists. TWAS would recognize and promote scientific excellence and be a leading organization dedicated to promoting global cooperation in science capacity building. In these ways, TWAS truly would be the voice of science for the South.

At the time, the voices of developing world scientists were not often heard at the global level. But in 1985, at TWAS's first general meeting, it was clear that the idea had momentum. The meeting was attended by UN

Secretary-General Javier Pérez de Cuéllar and Hans Blix, director-general of the International Atomic Energy Agency (IAEA). The theme of the meeting was telling: "South-South and South-North Cooperation in Sciences". Clearly, this signalled that new ambitions and new expectations would begin to shape science in the developing world.

A lofty vision

TWAS's vision of the future was based on an idea that we now take for granted, but which was, at the time, highly unconventional: Developing nations, by building their strength in science, engineering and technology, could begin to develop a sustainable strength that would support economic growth and human prosperity. It could help ease hunger, improve health and prepare communities for the risks posed by natural disasters. With this strength, developing nations could be more independent and less reliant on aid from developed nations.

The next year brought another auspicious development: Deng issued a personal invitation to Salam to hold the second TWAS General Meeting in Beijing. The meeting in September 1987 brought together 150 participants from 50 countries, plus 150 Chinese scientists and policy leaders. Among them were President Li Xiannian and Zhao Ziyang, secretary of the Central Committee of the Chinese Communist Party.

For me, that meeting marked an introduction to TWAS. It was there that I saw the Academy as one of the highest academic institutions in the developing world, and I learned that it was already very influential. That year, scientists from Brazil, India and Pakistan were among the first winners of the TWAS Prize. A fourth winner was the respected Chinese mathematician, Liao Shan Tao of Peking University. Liao's selection helped promote TWAS in the Chinese scientific community.

Some have described the 1987 Beijing meeting as a pivotal moment in China's history, in which the country began a scientific opening to the world. That is certainly true, but the meeting was significant for other reasons as well: It was an unmistakable signal not only from China, but from many nations in the developing world, that their latent potential in science and engineering was awakening. Science was now a priority.

Just a few years after its founding, TWAS already was established as a leader in the global movement toward science for development. Salam was the visionary and voice; Paolo Budinich helped to make it happen, securing critical support from the Italian government. Mohamed H.A. Hassan became the Academy's first executive director, and he would play a vital role in managing the Academy's affairs during a long period of growing influence.

Of course, even for great leaders and great organizations, history rarely proceeds in an even, ascending line. There are false starts and wrong turns, unexpected losses and setbacks. Like other organizations with high ambitions, TWAS has endured periods of challenge and uncertainty. But with strong leaders and sustained support from its Fellows and partners, TWAS has endured and grown. It has secured a place of leadership wherever scientists, policymakers, business executives and educators gather to discuss science in the developing world.

Today, TWAS has some 1,150 members from more than 90 countries. Its membership includes 15 Nobel laureates, top-level policymakers and presidential advisers, and the leaders of universities and nongovernmental organizations. We have more than 120 promising Young Affiliates. We have five regional offices and three associated organizations. We have dozens of partners from all over the world with whom we work in joint projects. The TWAS-Lenovo Science Prize is one of the most prestigious honours given to scientists in the developing world. During the past three decades, our advocacy, our training of young scientists, our research grants, and our policy advice and diplomacy have helped to strengthen science in dozens of nations and better the lives of millions of people. From high levels of government to modest research labs, we have partners and friends who cite TWAS as an inspiration.

A Voice for Science in the South tells the story of TWAS's birth and its rise to influence. It is an informal history, told by men and women who have been engaged directly with the Academy – from Founding Fellows and past presidents to scientists representing a new generation of leadership. Each of them has followed an extraordinary course to achieve individual success. But what emerges from their stories is that each has also felt an extraordinary responsibility – to science and to nations, to communities and to people worldwide. All of them have devoted a significant portion of their careers, and their lives, to TWAS.

Brazilian chemist José I. Vargas, who stepped in to lead TWAS shortly before Salam's death in 1996, articulated the idealism that they shared. "Science is not the only factor that determines the well-being of a society," Vargas writes in an essay for this volume. "But, as the Academy has illustrated throughout its history, science is certainly one of the key factors in the quest for a more equitable and prosperous world."

Mexican physicist Ana María Cetto Kramis has held high-level leadership posts for several major science organizations, including TWAS. "The unprecedented growth in scientific capabilities in developing countries is one of the most significant global developments of the past three decades," Cetto writes, "and TWAS deserves a great deal of credit for being an important player in this effort."

From these stories we are reminded anew of Salam's lofty vision for the role of science, of his extraordinary ability to transcend challenges – and of his seemingly irresistible powers of persuasion.

"Salam not only drew eminent scientists into the Academy's fold," says the Indian chemist C.N.R. Rao, a TWAS Founding Fellow and former president. "From the start, he also succeeded in convincing outstanding institutional partners to stand shoulder-to-shoulder with TWAS in their pursuit of shared goals."

"Salam was someone whom scientists, especially young scientists in the developing world, revered," adds Hassan. "As a preeminent researcher and an enthusiastic and untiring advocate for science, he was an esteemed figure – the personification of what a scientist could and should be."

For Brazilian mathematician Jacob Palis, who preceded me as the Academy's president, Salam's inspiration was embedded in a simple idea. Salam "often urged – indeed beseeched – colleagues and associates to 'think big'", Palis writes. "It was among his favourite expressions. He used it not only to convey his own convictions, but also to encourage others to join him in his cause."

The stories remind us as well of the remarkable solidarity and generosity of partners all over the world, whose encouragement and material support made TWAS's work possible. The government of Italy and its Ministry of Foreign Affairs have been friends and colleagues from the start, as has ICTP, through good times and bad. The Swedish International Development Cooperation Agency (Sida) has supported key programmes such as research grants and science diplomacy training for scientists in the developing world. The United Nations Educational, Scientific and Cultural Organization (UNESCO) has provided a steady administrative framework. Other governments, other agencies and other academies from many other nations similarly have honoured us with their support.

Finally, though, these pages contain the chronicles of the authors' own work and achievements. Their thoughtful recollections and reflections remind us of how important their leadership has been – not only to TWAS, but also to the advance of science throughout the developing world.

A focus on the future

During an anniversary year, it is natural to look at the past and assess the road we have travelled. This is an important process for any organization; it is essential to understand the past to be more effective in the future. For TWAS, many lessons can be learned by looking at Salam's writings, the Academy's early values and strategies, and the accomplishments of its past leaders and Fellows.

But TWAS, in its very DNA, is oriented to the future. This is especially important to remember as we work to evolve and innovate, and to remain relevant in a rapidly changing world.

These priorities are implicit in *A Voice for Science in the South.* The accomplished authors recall the past, and they rightfully respect it. But each also has valuable ideas about how TWAS should build on its success. Rao's words are instructive:

"TWAS would be wise to formulate new ways forward that will allow the Academy to remain as relevant – indeed as irreplaceable – in the future as it has been over the past few decades.... There is little doubt that the future of science in the developing world will be dramatically different than the past, largely characterized by new issues and new challenges. This is true about all facets of the scientific enterprise, including education, training, research, administration and science-based development and innovation.

"As a result, the fact that TWAS has been successful in the past – and continues to be successful today – is no guarantee that it will be successful in the future. It must keep up with the times."

Writers in this volume suggest a number of priorities for future work. I agree with their priorities, and I am confident that the TWAS Council does, too. Taken together, they comprise a roadmap for the years ahead:

1. Make TWAS broader, more inclusive – and stronger.

Abdus Salam famously said: "Scientific thought is the common heritage of mankind." Unfortunately, scientific capacity is distributed unevenly in our world – even among the developing nations. Therefore, if we want to nurture a new generation of scientists and engineers, we must reach further and search more diligently to find scientific excellence. This goal should be the foundation for our work in coming years. It has several important dimensions, each of which is important on its own.

- *Increase the number of women in the Academy.* This has been a priority for many years, and we are making progress. In 2014, we had many excellent nominees, and of the 46 new Fellows elected at our General Meeting in Oman, 10 were women. Still, among nearly 1,150 Fellows, women number just 10%, and it is essential that we increase that percentage through an effort sustained over many years. Further, we need to bring more women onto our committees and Council.
  Our interest, ultimately, is not to achieve certain numbers or percentages. Rather, we recognize that humanity faces extraordinary challenges, and we share the emerging consensus that it is essential to identify and nurture all of the world's scientific talent. TWAS's partner, the Organization for Women in Science for the Developing World (OWSD), plays a central role in helping us toward this goal. So does GenderInSITE, a promising new initiative that demonstrates how examining science and development through the lens of gender can provide deeper insights, more effective programmes and more sustainable outcomes. Both OWSD and GenderInSITE are hosted at TWAS's headquarters in Trieste, Italy.

- *Increase the number of younger scientists and engineers in the Academy.* Scientists are elected to TWAS because of their elite accomplishments, and often this recognition comes late in their careers. It is no surprise, then, that the average age of TWAS Fellows is 70 – well past retirement age in most nations. Many remain highly productive and valuable members of the Academy, but at the same time, we would benefit from the new experience and perspectives that can come with younger Fellows.

I was elected a TWAS Fellow in 1997, when I was in my mid-40s. This was a very high honour with practical value: TWAS provided me with a new platform that not only included more opportunities for scientific exchange and cooperation but also exciting prospects to build new friendships with colleagues from around the world.

Nurturing a new generation of researchers is central to TWAS's mission. Our support for young scientists takes many forms: PhD and postdoctoral fellowships, research grants and training opportunities. Our Young Affiliates programme allows some 125 highly promising scientists from the developing world to participate in our activities. But we should recognize that young scientists want substantive opportunities to apply their knowledge, creativity and idealism to important challenges. We should look for ways to support these ambitions.

- *Expand the Academy's geographical representation.* Today, TWAS Fellows represent some 90 nations, and that is admirable. But there are many countries where we have few members, or none at all. Often these are the countries that are most lagging in science and technology. Bringing honours to even one scientist in these countries can render lasting benefits. Not only does it give the scientist visibility and credibility; it also helps raise awareness of research in that country.In 2014, we mounted a campaign that brought nominations from a number of these under-represented nations. We elected our first member from Central African Republic, and a new member from Oman. We also elected our first member from Austria, a scientifically advanced nation that is generously hosting our 2015 General Meeting.

- *Expand the Academy's engagement with developed nations.* TWAS is known for building scientific strength in the South, and the developing world is rightfully proud of its progress in the past three decades. But from the beginning, we have had important partners in Europe, North

America and other areas of the developed world. Those links should be strengthened and expanded so that more benefits can be realized.
The innovation cultures of the developed world – including their R&D policies, elite schools and universities and their entrepreneurial climates – are in many cases highly effective. We could learn much from them through exchanges of students and faculty, joint research and other collaborative projects. Given its expertise and existing networks, TWAS is natural for the role of helping to build systematic South-North links among scientists, policy makers and institutions. To achieve that goal, we need to strengthen our own South-North links: continuing to elect excellent scientists from developed nations as TWAS Fellows, for example, and building new collaborations that encourage South-North exchanges of scientists (especially young scientists). We also need to utilize resources available from developed countries that can help to accelerate national, regional and global capacity building.
It also is important to expect that the benefits of these partnerships will, as much as possible, be shared equitably by the South and North. We have mutual interests, and we face profound common challenges. Just as developing nations are learning from the North, today there is much that the North can learn from the South's research and experience.

2. Address the growing gap between emerging countries and those that are less developed and least developed.

When we look at countries such as Brazil, India and China, or Mexico, Malaysia and South Africa, we see that TWAS's mission has contributed much to their success in the past 30 years. But this has had an unintended consequence: While a number of nations are now fully committed to using science and technology to help drive their economic development, other nations have made less progress, and many of the poorest nations have made very little. For example, Africa produces only a small fraction of the world's scientific publications – and most of those come from scientists in Egypt and South Africa.

TWAS and its key partners are already working to bridge this gap. We are seeking to select the very best scientists from under-represented countries as TWAS Fellows. The educational and research opportunities we offer with our partners help to train better scientists. Those who take advantage of our programmes often go on to become leaders in their institutions, and in their countries.

But TWAS should work with emerging countries on more systematic efforts to build capacity in lagging nations. One approach might involve the creation or expansion of short-term training programmes, such as those offered at the five CAS-TWAS Centres of Excellence. TWAS should also focus on policy areas that encourage closer cooperation between the less-developed countries, emerging nations, and the developed nations.

3. Shift more responsibilities to the five regional offices.

With a small staff and limited resources, the TWAS headquarters in Trieste has limited capacity to respond to the diverse and increasing needs of scientists and science institutions in the South. The TWAS regional offices have capably taken on a growing programmatic role, and that role should grow in the years to come.

These offices have more detailed knowledge of issues and needs in their regions, and of leaders and networks that could be partners in our mission. With this knowledge, they should play a bigger role in networking TWAS Fellows, in coordinating major programmes and activities, and in TWAS's regional outreach. Closer interaction and coordination among the regional offices are greatly needed. Ultimately, the regional offices should become effective executive arms of the TWAS secretariat in Trieste.

4. Develop a "think tank" orientation, enabling the Academy to pursue more vigorous engagement in science-related policies.

With the enormous expertise of its elected Fellows, and with its experience in policy at the national and international level, TWAS should be a global leader in providing policy advice. When nations or global bodies are shaping development goals or climate change policies, when they are devising programmes for women or educational systems, TWAS should be involved.

Our record of accomplishment is clear, and no other organization has TWAS's knowledge and credibility in providing the perspective of scientists and engineers from the South. A more ambitious engagement in the policy process would be of value not only to science in the developing world, but also to governments and policymaking bodies, educational institutions, funding agencies and other academies.

This effort could have several orientations: TWAS could seek to influence the science agenda in the developing world, providing non-partisan advice to national governments for the formulation of science and technology strategy, and in particular encouraging public investment in research and development. It could promote science for sustainability in the South, synthesizing research to aid decision-making in areas such as climate change, biodiversity loss, land degradation and natural disasters. And we could provide advice that supports efforts to build research capacity, based on our extensive experience with research grants, PhD education and faculty-exchange programmes.

This goal could also have training and communications components, with TWAS and its partners offering policymaking courses and workshops, and publishing high-level policy reports.

5. Effectively raise the funds needed to pursue these goals.

Throughout its history, TWAS has accomplished a great deal through a smart use of its limited resources. Ultimately, however, limited resources limit our capacity to serve as a strong advocate for science in the South. We need a creative and energetic approach to fundraising that seeks new partners at foundations, in business and among philanthropic organizations.

These five goals are highly ambitious; individually, and taken together, they represent our commitment to "thinking big". As we make progress in realizing these goals, we will raise TWAS's visibility and enhance our leadership. And, not least, we will contribute immeasurably to a new era of scientific advancement – in the developing world, and worldwide.

A time of historic challenge

Unlike TWAS's past presidents, I never knew Abdus Salam. I hope it may be said that his vision has been imparted to me through years of working with José Vargas, C.N.R. Rao and Jacob Palis – and with many other TWAS Fellows. But because I represent a new generation, I feel a special responsibility to understand Salam's ideas and assure that TWAS stays focused on the goals articulated at the founding three decades ago.

Thirty years is a long time, and so much has changed in our world. But I see parallels with that era and ours.

Back then, the world faced an urgent challenge: Poverty was a crushing daily reality for much of the world's population, and many nations were so poor that they were nearly paralysed. TWAS and its partners and allies showed that there was a way to break the paralysis, using science and technology for problem-solving and economic development, and to build hope.

Today, our challenge is no less urgent. The current world population of 7.3 billion is projected to increase by some 800 million just in the next decade; by 2050, the United Nations projects global population of 9.6 billion. Many of these people will be profoundly poor. The new Sustainable Development Goals call on us to eliminate poverty by 2030. At the same time, with economic growth, we can expect well over a billion new middle-class consumers. Therefore, we must grow more food and provide energy and clean water, sustainably. We must ensure health care. We must mitigate and arrest climate change, reduce pollution and aggressively conserve biodiversity.

This, then, is the value of *A Voice for Science in the South*: It will serve as a lasting reminder of the commitment that motivated Abdus Salam, Paolo Budinich and the others who founded the Academy and guided its work for the first 30 years. Though the world may change, their ideals remain constant. At the same time, this volume reminds us of the hard work and creativity that will be required to build on their legacy, so that TWAS remains an effective leader and advocate for science in the service of human progress.

Abdus Salam

Introduction

The vision of Abdus Salam

Daniel Schaffer

The World Academy of Sciences (TWAS), officially inaugurated in 1985, has been an insightful and influential voice for the promotion of science in the developing world during a period of momentous change. What began as an era marked by deep, intractable divides in scientific capacity between the developed and the developing worlds – the North and the South – has gradually but inexorably been transformed into an era characterized by broad advances in scientific capabilities across the globe. Amid all of the dramatic economic and social change that the world has experienced over the past three decades, the rise of science in the South may be the least appreciated but most consequential change of all.

That doesn't mean the transformation has been uniform. Some developing countries – for example, large countries with large populations – have achieved far greater progress than others. Nor is the journey complete. Scientific capabilities in the developed world still exceed those of the developing world, especially in such cutting-edge fields as biotechnology, materials science, and information and communication technologies. Developed countries, moreover, continue to lead the world in linking research discoveries to the global marketplace. Nevertheless, a convergence in capabilities is discernible, and if current trends continue, it is not difficult to foresee the disappearance of the North-South divide in science and technology over the course of the 21st century.

Statistics tell part of the story, and Abdus Salam employed them to great effect in speeches and articles during the period of TWAS's founding. In the early 1980s, just before the launch of TWAS, the developing world was home to more than 80% of the world's population but was responsible for less than 5% of the world's scientific publications. Scientists in the South produced an even more negligible share of international patents. Three decades ago, the North employed 3,600 scientists and engineers per million population. In contrast, the South employed less than 200 scientists and engineers per million population. There were virtually no universities in the developing world considered among the elite institutions of higher education and most had yet to institute either graduate or postgraduate programmes. In fact, universities in the developing world, especially those in Africa, had experienced a period of decline in the 1980s due to diminishing budgets, inadequate infrastructure, aging faculty and political unrest. In developing countries, annual investments in research and development (R&D) averaged 0.25% of gross domestic product (GDP). In developed countries, annual R&D investments averaged 2.5% of GDP. In real terms, this meant that the South spent about USD2 billion a year on science and technology and the North about USD100 billion.

Compare the situation then to the current state of affairs in science and technology in the developing world. According to the *2010 UNESCO Science Report* (the next report is due out in late 2015), scientists in developing countries account for 32% of the articles published in peer-reviewed scientific journals (although just seven developing countries – Argentina, Brazil, China, India, Iran, Mexico and Turkey – produce more than two-thirds of the total). Chinese scientists alone are responsible for 10.6% of the total, up from just 3% a decade before, and are now second only to the United States in global rankings. Similarly, nearly 40% of the world's researchers now reside in the developing world, with China again leading the way. International patents awarded to scientists and innovators in the South, although still lagging behind the North, continue to climb.

Meanwhile, the number of universities in developing countries has grown considerably, as has the quality of instruction. Graduate schools, especially those dedicated to the sciences, have increased in number and improved substantially in the quality of both their instruction and research. According to the Academic Ranking of World Universities, more than 50 universities in the South are currently listed among the world's top 500 institutions of higher education (although none are in the top 100). There is good reason to believe that this figure will grow in the years ahead as developing countries increase their investments in science education and, more generally, research and development. The latest research shows China's share of GDP devoted to science and technology at nearly 2%, and rising. Brazil's share stands at 1.3%, with India's at 0.9%.

While good governance and sound economic policies have received much of the credit for the progress that developing countries have made during the past three decades, investments in science and technology have clearly played a critical role in efforts to devise effective strategies for sustained growth. These efforts have been paying off. In 2013, developing countries and emerging economies accounted for more than 50% of the world's GDP. At the time of TWAS's creation, the figure stood at about 25%. China is now the world's second largest economy, and 12 developing countries and emerging economies now rank among the world's top 25 countries in terms of purchasing power parity.

All of these statistics are difficult to assemble with precision or complete confidence. Indeed one reason for the difficulty is that they are changing so fast. Yet the trends are as undeniable as they are inexorable. The world is experiencing a paradigm shift in the production of scientific knowledge that is relegating the North-South divide in science to a past era. It's a story that TWAS has not only witnessed but, even more significantly, has helped to write.

Guiding light

Abdus Salam's involvement in TWAS spanned 14 years – from his call for the creation of the Academy in 1981 to his resignation as TWAS president in 1995. Chronic illness had begun to curtail his efforts in the early 1990s before his death in 1996.

During this brief time, Salam masterfully led TWAS, transforming the Academy from a lofty concept and fragile organization into a significant player in global efforts to build scientific capacity and excellence in the South. Salam's legacy continues to this day. In a life brimming with great accomplishments, his Nobel Prize, the creation of the International Centre for Theoretical Physics (ICTP) and TWAS represent a lasting legacy.

Those who worked closely with Salam (a number of whom recount their interactions in the pages that follow) have never forgotten the experience.

Their collaboration with Salam is one of the highlights of their careers. It allowed them to nurture the friendship and trust of an iconic figure in international science and to be engaged in activities that reached far beyond their own personal ambitions. Their interactions with Salam, in short, helped them to make a difference in the world and for this they have been forever grateful. Nearly 20 years after his death, Salam remains a guiding light for scientists across the South – both as a mentor and inspiration. His remarkable achievements in life have been matched by his continual presence in the lives of others since his death.

The challenges that Salam faced in nurturing the creation and early development of TWAS were daunting.

He had to overcome profound scepticism among his colleagues, especially those in the developed world, who questioned the need for the Academy. They asked: Weren't there higher priorities and other international science institutions already in place to address the issues that the Academy proposed to deal with? Why spend precious resources on an institution that did not seem to address a critical need? Salam's sterling reputation, powers of persuasion and persistence served as a significant counterweight to these concerns.

He had to secure sufficient funding on a sustainable basis for an untested organization. He would attain this goal thanks to the generosity of the Italian government, which extended support to TWAS largely because of the success of his earlier brainchild, ICTP, and his central role in the Academy's creation.

He had to establish TWAS's credibility. He would achieve this by putting in place procedures that would ensure transparency and excellence in the election of TWAS members and the selection of Academy Fellows and prizewinners.

He had to quickly devise an array of effective programmes that would advance TWAS's agenda and have a perceptible impact on science in the developing world. He would accomplish this by launching awards, research grants, fellowships and lectureship programmes, all within a year of attaining funding for the Academy.

And Salam had to rapidly raise TWAS's public profile so that both the Academy's efforts and goals would gain global visibility and recognition. He would reach this goal, in part, by organizing high-level annual conferences, sponsored by the host governments. These conferences have become signature events for TWAS and, more broadly, science in the South.

Thanks in large measure to Salam's efforts, by the late 1980s TWAS had emerged as a highly respected organization in global efforts to build scientific capacity and excellence in the developing world. Within the first five years of its existence, the Academy became a key player on the stage of international science.

Halting progress

Salam was an eternal optimist – yet he was also realist. The tenor of the times in the 1980s, Salam warned, were at best only allowing for modest gains in scientific capacity building in the South. At the time of TWAS's creation, Salam cited just five developing countries that valued science: Argentina, Brazil, China, India and South Korea. For the rest, he said, science was largely "a marginal activity".

In fact, science in the South during this period remained in a dismal state. Salam, as a result, was deeply concerned about the prospects for improving conditions in the years ahead. While never falling into despair, he continued to be disheartened by the low level of scientific capacity that had been achieved. In a 1984 interview, he lamented: "Third World countries have yet to realize that basic science is a legitimate activity for them to support."

Throughout this period, he continued to beseech governments in the developing world to increase their investments in science and technology and to focus their efforts on the training of faculty and students and not simply on buildings and laboratories. He strongly urged oil-rich states in the developing world to establish collaborative sources of funding (foundations and multilateral financial agencies) that would help countries throughout the South improve their scientific and technological capabilities. More generally, he called for greater South-South cooperation in science and for the creation of global institutions that would facilitate such exchanges. He worried that the collapse of the Soviet Union would shift the focus of aid programmes from the South to the former Soviet republics in Eastern Europe. And he feared that the advent of new fields of science – for example, biotechnology, materials science and information and communication technologies – would widen the scientific gap between the North and South.

All these concerns were wrapped around ominous economic trends that foretold of possible difficult times ahead for science in the developing world. A steep decline in the price of commodities, together with rising levels of debt, were making it increasingly difficult for developing countries to find sufficient financial resources to invest in science and technology. In Salam's mind, the modest advances that had taken place in building scientific capacity in the 1960s and 1970s had largely stalled in the 1980s. He fretted that future progress would be placed increasingly at risk unless developing countries were able to reverse recent trends and accelerate the pace of scientific capacity building.

In many ways, Salam's writings in the late 1980s have remained remarkably current to this day. He continued to speak passionately about the critical role that science and technology play in economic development and poverty alleviation. He cautioned the developed world that gnawing gaps in scientific capabilities and economic strength between the North and South threatened world stability and peace. He emphasized the importance of supporting both women and young scientists as critical aspects of the global scientific agenda. He noted that such seminal issues as food, water and energy security, the spread of infectious diseases and climate change posed critical global challenges, and he asserted that none of these issues could be solved without scientific knowledge – nor without the active engagement of the developing world. And he maintained that scientists, especially those in the developing world, should not sit on the sidelines but had an obligation to serve as strong advocates for the promotion of science.

Changes in place

Nearly two decades after his death, Salam's broad list of concerns continues to align with the current agenda for science in the South. What has changed, however, is the context in which these challenges exist. The broadening spheres of scientific capacity and excellence that are currently sweeping across the developing world simply were not present at the time of TWAS's creation.

What Salam earnestly hoped for, but could not have fully anticipated during the concluding years of his life, was the dramatic increase in scientific capacity among developing countries that would begin in earnest in the 1990s and continue to this day. He could not have anticipated the profound impact of information and communication technology (ICT) on science in the South. (In the early 1990s, TWAS and ICTP were donating more than 45,000 printed books, journals and conference proceedings to libraries across the developing world each year.) Moreover, he could not fully envision the impact that the end of the Cold War would have on world diplomacy and global aid programmes. (In the final years of his life, Salam fervently called for a peace dividend that would shift global investments from defence and nuclear weaponry to sustainable economic development, especially in the developing world. He also called on the North to establish a Marshall Plan for the South.) And, perhaps most importantly, he could not have foreseen that scientific and technological capacity building would become a cornerstone of economic development strategies throughout much the developing world, in countries ranging from Mexico to Qatar and from Malaysia to South Africa.

In many respects, what Salam had valiantly fought for has become a reality: The North-South gap in scientific and technologically capacity has narrowed considerably over the past two decades, increasingly blurring the lines between developed countries and a growing number of rapidly advancing developing countries.

A question of time

TWAS has been an active participant in this historic evolution and, despite its small size and budget, the Academy deserves credit for helping to raise global awareness and transforming attitudes about the importance of science and technology for development.

TWAS was promoting scientific and technological capacity building in the developing world at a time when few other institutions or countries were, and it was one of the first organizations to consistently call for South-South cooperation in science. The Academy established one of the first research grant programmes for scientists in the South. It also launched one of the first awards programmes devoted exclusively to scientists from the developing world. With the help of the South's most scientifically advanced nations, the Academy created one of the world's largest doctoral and postdoctoral fellowship programmes for early-career scientists in developing countries. Through its general conferences, the Academy helped to foster closer ties between the scientific and policy communities in the developing world, playing an important role in integrating science and technology into broader economic development strategies.

Over the past decade, TWAS has significantly broadened it programmes to assist young scientists, and it has taken measures, through its regional offices, to decentralize its activities in recognition of the increasing strength of science across the developing world. Thanks to financial support from Lenovo, China's international technology giant (it is the world's largest PC company), TWAS now oversees the USD100,000 TWAS-Lenovo Science Prize, given annually to the South's most accomplished scientists. To reflect the rapidly changing state of science in the South and its growing presence in the international scientific community, it has even changed its name twice in the past decade – from its original name, the 'Third World Academy of Science', to 'the Academy of Sciences for the Developing World', to 'The World Academy of Sciences for the advancement of science in developing countries'.

Having noted earlier that no one, not even Salam, could have projected the rapid evolution in the global scientific landscape that has taken place over the past few decades, it would seem foolhardy to project the changes that lie ahead. Yet, this much is certain: The world of science that was firmly in place at the time of TWAS's launch and during the first decade of the Academy's existence is fading into history, and many of the fundamental issues that TWAS has sought to address are increasingly playing out in an vastly different global context.

The era of scientific capacity building, as it has been defined since the end of World War II, is rapidly coming to a close. A world sharply defined by 'have' and 'have-not' nations – a world divided between North and South – no longer reflects either the state of global science or the state of the global economy. A more complex and intricate global mosaic is rapidly displacing the two-dimensional world characterized by a scientifically superior North and a scientifically lagging South. Knowledge and wealth are emerging in many more places. Impoverished and marginalized populations, in turn, are increasingly being consigned to 'left-behind' portions of society found in both the developing and developed worlds. Overlaying all of these historic changes is the rise of complex, science-related global challenges – for example, climate change, food and water scarcities, and the spread of infectious diseases – that can only be met by a global response.

Such developments will dramatically alter where wealth resides and how science is conducted on an international scale. This, in turn, will create both challenges and opportunities for international organizations such as TWAS.

Thanks to the impeccable reputation of its membership, the effectiveness of its programmes and the high level of respect and visibility it enjoys in the international scientific community, TWAS has never been in a stronger position to have its voice heard. Yet, in light of the dramatic progress that has been made in scientific capacity in the South, the Academy will need to reassess its strategy in the years ahead.

Success: evolving definitions

As many of the contributors to this volume note in the essays that follow, TWAS in the years ahead cannot rely solely on the same agenda it has pursued in the past three decades; it must grow and evolve. Yet it will not be easy to recalibrate its strategies to meet the future challenges of science and economic development in a world that has experienced such fundamental change.

So, on the occasion of its 30th anniversary, even as it celebrates its past accomplishments, TWAS is also finding it necessary to re-examine its future role within the global scientific community. Among the major challenges likely to be faced in the years ahead are:

- *South to South.* The North-South gap in scientific capacity has been at the centre of the Academy's concerns since its inception. But over the past two decades, another critical gap has emerged: a South-South gap between scientifically proficient countries and those that are scientifically lagging. Scientists from Brazil, China, India and a few other large developing countries dominate both the Academy's membership and its roster of grant, fellowship and prize recipients. In recognition of this challenge, the Academy has identified 81 scientifically lagging countries, located largely in sub-Saharan Africa and countries with predominantly Muslim populations, and established a grants programme designed exclusively for researchers in these countries. But it will have to take even more rigorous steps to ensure that its activities, in large measure, benefit countries with the greatest needs. Without aggressively reaching out to scientifically lagging countries, TWAS's goal to help build adequate levels of scientific capacity in all countries cannot – and will not – be met. Should the South-South gap in science be TWAS's primary concern? If so, how should the Academy's agenda be adjusted to meet this challenge? As the South becomes increasingly varied in its scientific and economic makeup, how can the Academy devise a strategy that satisfies the needs of its increasingly diverse constituents?

- *South to North.* Even as it has focused intently on the challenges faced by scientists in South, TWAS always kept its doors open to the North. The Academy's inaugural conference in 1985, 'South-South and South-North Cooperation in Sciences', was devoted to both hemispheres. Most of its early members were educated in the North, and nearly half lived and worked in developed countries. Even today, about 15% of the Academy's members reside in the developed world. TWAS, moreover, has always welcomed opportunities to participate in international activities and programmes where scientists from the North largely determined and implemented the agenda. But as TWAS's reputation in the global scientific community has grown – and as science in the South has become stronger – the Academy's relationship with the Northern scientific community has changed. The North now eagerly seeks the Academy's involvement in international global initiatives, increasingly embracing TWAS as an equal, not junior, partner in such ventures. How should the Academy use its elevated status to help shape the global scientific agenda? How can the Academy make certain that such activities do not come at the expense of its primary goal to build scientific capacity in the South? Are there ways to advance both goals – North-

South and South-South capacity building – simultaneously? Can TWAS serve as a valuable bridge between the developing and developed worlds? The science diplomacy partnership between TWAS and the American Association for the Advancement of Science (AAAS) is a welcome initiative that seeks to address the challenges and opportunities presented by the Academy's growing presence in the international scientific community. What more can – and should – be done?

- *Nation to nation.* In the past, when devising strategies for scientific capacity building and economic prosperity, international organizations such as TWAS focused on distinctions between nations. Increasingly, however, scientific and economic gaps are opening up within nations. It could be argued that scientists working in state-of-the-art facilities in Brazil, China, India and a growing number of developing countries have more in common with colleagues in the United States, Europe and Japan than they do with their fellow citizens. In a similar vein, there is evidence that rural and remote areas in developing countries have failed to benefit from advances in science to the same degree as their urban counterparts. Does TWAS have an obligation to examine and respond to the sub-national divides that are now unfolding, or is this a challenge that should be left largely to the nations themselves? What measures, if any, should the Academy take to address the domestic gaps in science and economic well-being that have become so prevalent in the South (and the North as well)?

- *Expanding the playing field.* Salam had great respect and admiration for the United Nations (UN). He firmly believed that the UN provided the best forum for guaranteeing that the concerns of the developing world would be heard and addressed. Not only has the UN (most notably, UNESCO) provided valuable credentialing for the Academy, but it also has been an indispensable partner in efforts to advance the Academy's goals. TWAS could not have attained its current level of success and prestige without this affiliation. However, as the global scientific community becomes more diverse and decentralized, TWAS will likely need to forge greater collaboration with national and subnational players in both the public and private sectors. Its doctorate and post-doctorate fellowship programmes, conducted in collaboration with governments and institutions across the developing world, represent an excellent example of the Academy's outreach to national scientific networks in the South. Should the Academy increase its efforts even further to work not only with international and national entities but also non-governmental organizations and private-sector enterprises to advance its agenda?

- *Sustaining progress.* More than a concrete goal, scientific capacity and excellence is an open-ended process that must be transmitted from one generation to the next. Given the South's growing scientific capabilities, the future agenda of TWAS may well come to encompass both "building" and "sustaining" scientific capacity and excellence. To help fuel progress for generations to come, should the Academy intensify its efforts to promote scientific education and training, not only at the tertiary level but also at the secondary and even primary levels? Should TWAS become more engaged in policy discussions designed to strengthen the role of science in global efforts to expand economic well-being, improve public health and safeguard the environment? Should it take a leading position in examining such critical global issues as climate change, species loss and food security, drawing on its unique position in international science and its wide-ranging networks, especially in the developing world, to present unique perspectives on critical global issues? As several contributors to this volume suggest, such efforts will require the Academy to give additional responsibilities to its regional offices, especially for the administration of its graduate and postgraduate programmes. A devolution of authority would free the TWAS secretariat to pursue a broader policy agenda to sustain the progress that has been made. What measures, then, should be taken to ensure that the regional offices have the resources and requisite personnel to shoulder these responsibilities?

A new world

At the time of TWAS's creation, the reasons for the developing world's inability to build adequate scientific capacity were well known – and, even more significantly, seemed to be firmly entrenched. Paltry spending, inadequate infrastructure and a chronic sense of isolation experienced by scientists all contributed to the dismal state of science in the South. Driving all of these shortcomings was a lack of commitment to science, most notably on the part of governments.

But that world is receding into history, and a new world is rising in its place. This new world presents challenges that are equally daunting but, in many ways, profoundly different than those that dominated the global scientific agenda during the last half of the 20th century.

At the start of TWAS's fourth decade, examining the past with a keen eye to the future may be the best way to celebrate and commemorate TWAS's 30th anniversary and to make sure that the Academy continues to be an essential player in global efforts to build a more equitable and peaceful world. That was the ultimate goal that drove Abdus Salam's agenda in the creation and early development of TWAS and that has been the sentiment of all those who have played a prominent role in TWAS's success ever since.

To continue to succeed, the Academy must continue to evolve. While its lofty goals are likely to remain largely the same, the tactics to achieve those goals will need to be altered to address the new realities that the Academy – and science in the developing world – now face. To move forward, TWAS will need to draw upon the traits that have propelled the Academy ahead since its inception: courage, conviction, innovation, boldness and relentless determination. But it will need to apply these traits in a dramatically transformed global context that the Academy itself has helped to create.

That is the key message found in the pages that follow – a message that resonates among the men and women of many nations who have been among the Academy's strongest advocates. It is a message that celebrates and honours the Academy's past while setting the stage for an equally successful future. It represents a way to congratulate and honour TWAS for its remarkable past and current impact while reminding the Academy that its work is never done.

Present in the world

José I. Vargas

José I. Vargas was born in 1928 in Paracatú, Minas Gerais, a commercial centre of less than 100,000 people in south-central Brazil. The region was once dominated by gold mining and later by agriculture and livestock. Today it is enjoying an increasingly diverse economy. He attended primary school in his hometown. His family subsequently moved to Belo Horizonte, the state capital of Minas Gerais, where he graduated from secondary school and earned an undergraduate degree from the Federal University of Minas Gerais (UFMG) in 1952.

Upon graduation, Vargas taught secondary school for a year. He then became a postgraduate student in physics at the Technological Institute of Aeronautics (ITA), where he was hired as a researcher. In 1954, he attended a nuclear chemistry course, in Concepción, Chile, organized by Cambridge University. He was accepted to Cambridge University the following year and received a PhD in chemistry in 1959.

With his advanced degree in hand, Vargas returned to Brazil to resume his academic career at UFMG. In 1962, he was named a Brazilian commissioner and alternate delegate to the board of governors at the International Atomic Energy Agency (IAEA) in Vienna, Austria. This marked his initial experience as a diplomat. Following a military coup in Brazil in 1964, Vargas was relieved of his post. He accepted a position with the French Atomic Energy Commission's Centre for Nuclear Studies in Grenoble, where he remained for six years.

Upon returning to Brazil in 1972, Vargas was appointed to a series of research and administrative posts, reflecting his broad-ranging talents for science and administration. He served as an advisor to the Research and

Projects Financing Agency (FINEP); a member of the plenary board of the National Council of Research (CNPq); state secretary for science and technology for Minas Gerais; and federal secretary for industrial technology. In 1992, he was appointed the Brazilian minister of science and technology, serving in that position until 1998.

Vargas also has held a variety of high-level positions in international organizations, most notably in UNESCO, where he was Brazil's chief delegate to the executive board, chairman, and later ambassador. In addition, he has chaired the International Labour Organization's (ILO) Science and Technology Committee and served on the board of the United Nations University's Institute of Advanced Studies (UNU-IAS).

He was elected a member of TWAS in 1988 and became vice-president in 1992. When Abdus Salam's poor health forced him to step aside from leading the Academy, Vargas was appointed TWAS's interim president in January 1995; in September 1995, he chaired the TWAS meeting in Nigeria; and in November 1996, just days after Salam died, he was officially chosen to succeed Salam at the TWAS meeting in Trieste.

Vargas would remain the Academy's president until May 2000. His tenure was highlighted by significant progress in securing permanent funding from the Italian government, a renewed effort to expand the Academy's endowment, a growing presence of TWAS in South America, and a heightened profile of the Academy in international policy and economic development circles. During his tenure, TWAS became an increasingly strong voice for the promotion of science-based development in the South.

When Vargas took the reins of TWAS, there were serious questions about whether the Academy could survive the loss of Salam. When Vargas left the presidency, few doubted the Academy's future as a prominent player in international science.

Vargas has won many honours during his long and illustrious career, including the IBM Science and Technology Prize, the International Cooperation in Science and Technology Prize in China, the Leloir Medal in Argentina, the French Legion of Honour, and the Knight Commander of the British Empire. He serves on the boards of numerous institutions, both nationally and internationally, and is a member of the Brazilian Academy of Sciences, the Buenos Aires National Academy of Sciences and the European Academy of Arts, Sciences and Humanities.

In the following essay, Vargas describes how support for the creation of TWAS emerged from broad currents in global science during the post-World War II era. These currents helped to underscore the importance of science to society and the need for developing countries to build their own scientific capacity in a world powered by knowledge.

***

In the final analysis, TWAS's existence is largely due to the brilliance of Abdus Salam, the Academy's founding president, leading inspirational figure and primary strategist. Salam's achievements beyond his personal triumphs as a world-renowned scientist and Nobel laureate – enshrined in the International Centre for Theoretical Physics (ICTP) and The World Academy of Sciences (TWAS) – prove that extraordinary individuals can make a difference. In a world defined by endless complexities and challenges, genius and determination do indeed matter.

Yet it is also true that even concepts and causes conceived by genius arise within a larger social context. This helps to explain why unique ideas sometimes resonate among large numbers of people to find their way to general acceptance and success... and why sometimes they do not. Few doubt that timing and circumstance are key factors in success, even for those endowed with special gifts.

I would like to explore this larger context in the opening pages of this essay. Such a contextual discussion, I believe, helps to shed light on why the timeless and universal principles that TWAS embodies ultimately found expression at a particular time and in a particular place.

In the age of the atom

I believe that two broad events – one global and virtually unmatched in its historical significance, and another more immediate and mundane – are critical to understanding the forces that gave rise to TWAS in the early 1980s.

The advent of the nuclear age – marked by the atomic bombs dropped on Hiroshima and Nagasaki, Japan – has been rightfully viewed as a seminal development in the history of humankind.

Few human events since the beginning of time have been as instrumental in shaping the course of history. For some, the most important aspect of the atomic bomb is that it ensured an Allied victory and ended World War II. For others, it became the defining element in geopolitical relationships and tensions that marked the Cold War era that followed. For still others, by virtue of the intellectual brilliance that nuclear research represented and its profound impact on world affairs, it ushered in a golden age of science in developed countries, especially in the United States, with physicists migrating from across the globe to lead the way.

What is not often mentioned in discussions of the advent of the nuclear age is the impact that it had in science in the developing world. Support for a generation of scientists in the South – though small in number – was based, in part, on the dramatic events that took place during World War II and that shaped the contours of the subsequent post-war period.

Educated as a nuclear chemist, I was part of that generation. Abdus Salam, educated as a high-energy nuclear physicist, was part of that generation, too – as were many of the scientists who were first invited to ICTP and later became active in TWAS. Physicists dominated both organizations at their inception and for many years thereafter.

The theoretical research that gave rise to the nuclear age was an example of science having the capability to change the world forever. It was also an example of how science, at least in the conceptual realm, could level the playing field for those working in countries with poor laboratory facilities. It was, after all, not the quality of the laboratories that propelled the ideas of theoretical physics and related fields forward; rather, it was the quality of the thinking of the researchers.

Nuclear science, moreover, was an intellectual pursuit that government officials throughout the world could appreciate and support – East and West, North and South. It may have been ironic and somewhat disconcerting, but nuclear research made science relevant to decision-makers in ways that other fields – no matter how significant to social well-being – did not.

The Atoms for Peace project, launched in 1952 by the United States, became one of the cornerstones of the International Atomic Energy Agency (IAEA), which was officially launched in 1957. IAEA activities, in turn, broadened global interest in nuclear science. Of special interest to developing countries was the notion that nuclear energy could render electricity "too cheap to measure". This created the tantalizing prospect that poor countries could potentially have access to virtually cost-free energy.

Thus, despite all the fear that nuclear science engendered, it also served to elevate the interest of developing countries in science and created an important reason for countries to invest in it during the immediate post-World War II period. This was especially true in large developing countries such as China, India and my own country, Brazil.

Science when needed

The lustre of science in the developing world would dull significantly beginning in the 1960s when an alternative paradigm for development would emerge. This new paradigm, which would gain strength in the 1970s and 1980s, was based on the notion that science was a luxury that developing countries could not afford and that technology simply could be purchased from what one proponent called the "supermarket of science" in developed countries as the need arose. It would be better, the argument went, for developing countries to concentrate on more immediate challenges and focus their attention on soft, small-bore scientific challenges, such as ensuring adequate and reliable supplies of water, food and energy. The pathways to improved economic and social prosperity in developing countries, proponents of this paradigm contended, could be navigated largely without indigenous science since solutions to these challenges had already been devised elsewhere and could simply be adopted.

Salam, along with a few colleagues from the developing world, never bought into this proposition. In fact, from the 1960s onward he remained one of the few voices continuing to contend that developing countries needed to build their own scientific capacity regardless of the technologies that they might be able to purchase off-the-shelf from developed countries. He vehemently argued that "there could be no applied science without science" and that scientists in developing countries must at least be competent enough to "read the labels" accompanying the products and services that their countries were purchasing.

While Salam's vision of the future of science in the South may have been lofty and idealistic, his feet were always planted firmly on the ground. He was a person of immense practicality (yes, a theoretical physicist with a practical bent) who recognized the world for what it was – even as he sought to uproot its norms and plant the seeds for dramatic global change.

False hopes, new strategies

As a result, it was another event that took place in the late 1970s, some two decades after he had begun his life-long campaign to build scientific capacity in the South, that I think helped to spur Salam's thinking on the need to create an academy for scientists from developing countries. The event was the UN International Conference on Science and Technology for Development (UNCSTD), held in Vienna, Austria, in 1979.

The Group of 77, the primary advocacy group for the 120 developing countries that belong to the UN, organized UNCSTD. The conference's goals were nothing less than to devise a broad strategy for poverty reduction and wealth creation by drawing on the unprecedented opportunities afforded by science and technology.

I was elected vice-president of the conference, which was preceded by a lengthy period of organization marked by pre-conference policy meetings in some 60 countries. I remember the grand vision that drove the conference's agenda: a call for USD20 billion in investments in science and technology in developing countries designed to place them on a track for sustainable development; a belief that if this investment was made, the conference's vision could be fulfilled in 20 years; and the establishment of a secretariat at the UN headquarters in New York City to carry out this lofty agenda.

The conference, however, produced only meagre results. The enthusiasm and hope fostered in the run-up to the event quickly deflated once the conference began. Even the modest plan outlined in the Vienna Programme of Action, presented at the conclusion of the conference, proved difficult to implement in the years that followed.

While UNCSTD enabled the North and South to air their differences over the role of science and technology in development, it did not provide a clear and compelling blueprint for sustainable development based on science and technology. One of the few concrete recommendations of the conference, calling for the creation of a UN "interim fund for science and technology for development" totalling USD250 million for 1980-1981 as a down payment for its ambitious multi-billion programme, never reached fruition. Voluntary contributions trickled in at a distressingly slow pace, ultimately plateauing at just USD40 million.

UNCSTD, which began with high expectations but ended in acute disappointment, convinced Salam that the North would never provide the South with the level of assistance it needed to implement a realistic plan for building adequate scientific capacity to curb poverty and increase wealth. In light of this experience, Salam came to the conclusion that developing countries would only be able to attain these goals through their own initiative.

Science requires scientists

As a corollary to this viewpoint, not surprisingly, Salam also believed that a strategy based on self-reliance would ultimately require a cadre of well-trained scientists living and working in their native countries. Given this reality, an international academy of sciences dedicated to scientists in the developing world would help chart a more effective path for development.

The immediate outcome of the Vienna conference provided ample reason to be disillusioned. The lack of follow-up in the years following the conference only added to the sense that the current global strategy for science-based development in the developing world, which depended disproportionately on the goodwill and benevolence of the developed world, would not be sufficient to do the job.

Salam spoke about his deep disappointment with UNCSTD in speeches and presentations that he gave over the course of the 1980s and 1990s. The conference's shortcomings, in my estimation, were a significant factor in prompting his call for the creation of TWAS in 1981 and then in devising a broad and ultimately successful strategy for its creation over the next two years – culminating in the Academy's launch in 1983.

Callings in science and diplomacy

My journey as a scientist in the developing world, seeking to build a career in the post World War II era, parallelled the journey of many of my contemporaries in Brazil – and I suspect many other young scientists in developing countries.

I was born and raised in Paracatú, a small commercial centre in a vast agricultural and livestock region in the state of Minas Gerais. In the 18th and 19th centuries, the region had been the focal point of colonial gold mining that brought untold riches to Portugal. I earned an undergraduate degree in chemistry at Federal University of Minas Gerais (UFMG) in Belo Horizonte in 1952. Belo Horizonte, the capital of Minas Gerais, was – and remains – at the heart of the nation's third most populous metropolitan area. Only the metropolitan areas of São Paulo and Rio de Janeiro are larger.

Upon graduation, I began teaching secondary school physics in Belo Horizonte. I had the good fortune to be selected to attend a summer school for teachers organized by the Technological Institute of Aeronautics (ITA) near São Paulo. At the time, ITA was the sole postgraduate school in Brazil as well as the nation's premier engineering school. It has held the latter distinction to this day.

Several of those teaching at the summer school, including Gleb Wataghin and Giuseppe Occhialini, were world-class physicists who had been recruited from the University of São Paulo, which was nearby. They had fled Europe in the 1930s during the rise of Fascism and Nazism to become the founders of Brazil's modern physics research community. The renowned US American physicists Richard Feynman and David Bohm were also there, the latter a refugee from the McCarthy era's politically motivated assaults on his loyalty in the early 1950s. Needless to say, for a secondary school teacher to be part of such a rarefied research environment was an eye-opening experience.

Physics research at ITA, which focused on the complexity of the nucleus and cosmic showers, had a profound impact on the future of Brazil's scientific community. The exploration of mesons – subatomic particles that are smaller than protons and neutrons – held the promise that nuclear energy could be generated cheaply. This, in turn, attracted the attention of the government in Brazil, which eagerly sought to increase its investment in science as part of a larger effort to improve the country's economic performance.

The meson's too-brief half-life made it impossible to use it as an energy source. The particle's burst of energy is too fleeting to be harnessed. Nevertheless, its brief promise as a possible source of cheap energy convinced the government to substantially raise the level of funding for the National Council for Research (CNPq). That, in turn, helped to strengthen CNPq's role as one of the primary pillars of the nation's research infrastructure. In a similar vein, the co-discovery of the meson by Cesar Lattes, a young researcher at the University of São Paulo, prompted the creation of the Brazilian Centre for Physics Research in Rio de Janeiro, which soon became another node for collaborative research in the nation's expanding physics community. (Lattes was elected a TWAS Fellow in 1988.)

Following my attendance at ITA's summer school, I was recruited to be a researcher there. Two years later, I was accepted to Cambridge University in the United Kingdom, where I earned a PhD in chemistry in 1959. My research was closely aligned to physics and, over the course of my studies, I collaborated with physicists as much as I did with chemists. In fact, my hybrid research activities led me to become Brazil's first nuclear chemist with a PhD. My years as a doctoral student, moreover, were marked by my first extended stay in Europe.

During my studies, I also had the opportunity to meet Abdus Salam for the first time when Jayme Tiomno, an acclaimed Brazilian physicist and later a TWAS fellow, invited me to attend a lecture Niels Bohr was giving at Imperial College in London, where Salam was a professor. My first encounter with Salam – if you can call it that – amounted to being introduced to him by Tiomno in a large lecture hall.

In 1959, after earning my doctorate, I returned to Brazil, to become a professor of physical and advanced chemistry at UFMG. Despite the political turmoil engulfing the country, the times offered unexpected opportunities for young scientists. The government, whether ruled by democratic leaders or military authorities, was showing increasing interest for what science, especially physics and chemistry, could do for society.

In 1957, Brazil received its first nuclear research reactor from the United States under the Atoms for Peace programme. The reactor was housed at the Institute of Atomic Energy in São Paulo. Two years later, Brazil received a second reactor. It was placed at UFMG's Institute of Radioactive Research in Belo Horizonte. In 1962, Brazil completed the construction of its first indigenous reactor, located at the Nuclear Engineering Institute in Rio de Janeiro.

In 1964, following the military's overthrow of democratically elected officials, the government accelerated its efforts to develop an indigenous nuclear energy industry. The goal was to acquire capabilities for managing the complete nuclear fuel production cycle. In effect, Brazil was placed on a fast track to become a full-fledged member of the world's elite nuclear club.

In 1962, when I was just 26 years old, I was one of four people nominated to represent Brazil on the governing board of the International Atomic Energy Agency (IAEA) in Vienna. Until that point, my 'diplomatic career' had consisted of travelling overseas to obtain an advanced degree and participating in studies of Brazil's emerging nuclear energy programme for the National Nuclear Energy Commission. Other than that, I had worked one year as secondary school teacher and two years as a university instructor. I had interacted with scientific colleagues, conducted research and published articles in peer-reviewed journals. I may have shown promise as a scientist. But I had done virtually nothing to indicate that I had either the aptitude or skills to be a diplomat. My experience, in short, was no match for the job I was given.

But that was the nature of things in those days. Young, highly educated people in Brazil in the 1950s and 1960s were given challenging opportunities and placed in leadership positions that subsequent generations of Brazilians could only dream of replicating during the early stages of their careers. Living in a unique moment, we were the material – the raw material – that our countries had to work with. They turned to us because that was all they had.

As a Brazilian delegate to the IAEA, I was introduced to a new world that examined science not as a tightly bound discipline and field of study, largely confined to a community of experts, but also as an extraordinary societal tool that exerted broad and lasting impacts on society.

Both of these worlds of science were international in scope. Yet, despite the large swatches of common ground they occupied, they were worlds largely led by individuals with different skill sets and different sensibilities. One of Salam's great gifts was to be able to operate successfully at the highest levels in each – and to serve as a bridge between the two.

Towards Trieste

At the time of my arrival in Vienna, Salam was a delegate to the IAEA for Pakistan. What is less well known is that Paolo Budinich (TWAS Fellow 1992), a physics professor at the University of Trieste and a vocal and relentless proponent for transforming Trieste into an international centre for science, was a delegate to the IAEA for Italy.

Salam and Budinich had formed an unusual alliance in support of the creation of an international centre for theoretical physics dedicated to providing training and research opportunities for scientists from the developing world while also serving as a 'free zone' for East-West exchanges on nuclear research. Salam focused on the value that such a centre would have for physicists and mathematicians from countries where poor working conditions made scientific careers virtually impossible to pursue. Budinich, in turn, emphasized the value that such a centre would have for Italy and the city of Trieste, particularly as a tension-free meeting place for scientists from the East and West.

Salam and Budinich thus forged a powerful alliance that conveyed how the centre could simultaneously serve the global scientific community and the interests of Italy. Their vision was driven not only by high ideals of global cooperation and progress but also by the down-to-earth realities of geopolitical challenges and opportunities.

My tenure at the IAEA as a delegate from Brazil corresponded to the time that the Agency's member states, following several years of discussion, voted on whether to authorize the creation of what ultimately became ICTP.

The outcome was by no means assured. Wealthier countries were doubtful about the need for such an organizations. They contended that some of the same goals could be achieved more economically by sponsoring fellowships and training activities in existing institutions. Others questioned whether responsibility for such an institution fell under the Agency's mandate. With strong support from representatives from developing countries, however, the proposal passed, with Trieste being chosen as the site for the nascent institution.

I am happy to say that I voted in favour of the Centre, which has become one of the world's foremost international centres for scientific training and research and a cornerstone of global efforts to promote scientific capacity building and collaboration.

ICTP, as many observers know, also helped to nurture TWAS, especially during its early years. It gave the Academy the support it needed (in the form of personnel, office space and global contacts) to gain sufficient traction for success. The two organizations remain close, continuing to operate as key bulwarks in the campaign to build scientific capacity and excellence in the South.

The way home

In 1964, I was expelled from my position at the IAEA by Brazil's military, which had assumed political power following a coup. My activism as a university student made me suspect within government circles, prompting the newly installed government to relieve me of my duties. My ability to work freely was also circumscribed when I returned home. As a result, in 1966, I decided to accept an offer from the French Atomic Energy Commission's Centre for Nuclear Studies in Grenoble, where I remained for six years, pursuing an active research agenda.

I returned to Brazil in 1972. Military authorities were still in control. However, recognizing the potential value of science, they had 'liberalized' their policies to allow scientists to conduct research in a largely unfettered manner, especially in areas that the government viewed as beneficial to the economy.

When a society is not free, no one is free. Yet, to a degree, scientists in Brazil in the 1970s and early 1980s were granted the latitude they needed to pursue productive careers in ways that other citizens could not.

I was given a teaching and research position with UFMG, where I had previously been a student and professor. In 1973, I was also asked to be part of a commission that had been mandated to examine Brazil's nuclear energy programme. Specifically, the commission was asked to explore what measures should be taken to ensure there would be sufficient numbers of well-trained personnel to meet the nation's future research needs in what was perceived to be a critical field of study. For me, this marked the first of a series of appointments to increasingly high-level posts in state and federal government, culminating with being named Brazil's minister of science and technology in 1992. My initial appointments were due in no small measure to my close relationship with Aureliano Chaves, the governor of the state of Minas Gerais.

In the early 1970s, I was also asked to chair the state of Minas Gerais planning commission. This gave me oversight responsibilities for such critical sectors of the state's economy as the environment and science and technology. In addition, I was appointed to the plenary board of the National Research Council (CNPq), Brazil's key organization for the development of science and technology at the federal level.

In 1978, Chaves became the nation's vice-president. I followed him to the capital city of Brasilia when I was named secretary of industry and technology. There I oversaw the launch of Brazil's ethanol project, which has since become one of the nation's most successful science-based initiatives for economic growth.

Today, Brazil is the world leader in ethanol technology and production. More than 85% of the nation's vehicular fleet is equipped with 'flex' engines capable of operating with either ethanol or conventional fossil fuels. Refineries across the nation, moreover, produce gasoline with 20% ethanol content. In total, ethanol replaces the equivalent of more than 250,000 barrels of oil each day.

As my responsibilities within Brazil's science policy community increased, so too did my involvement in international organizations, especially those organizations dedicated to economic and scientific development.

During the 1970s, for example, I participated in negotiations with the World Bank to secure an innovative loan for scientific and technological development. Brazil became the second nation to obtain a loan for this purpose; South Korea was the first. I was also named vice-president of the organizing committee for the 1979 UN International Conference on Science and Technology for Development, which – as mentioned earlier – ultimately failed to fulfil its ambitious but flawed agenda. And I served as chairperson of the Advanced Committee on Science and Technology at the International Labour Organization (ILO), a UN organization dedicated to monitoring and improving conditions for workers worldwide.

In the mid-1980s, I left government and returned to academia at UFMG. At the time, I also became actively involved in the Brazilian Academy of Sciences, where I was vice-president for more than a decade. I remained active in science policy issues both at the national and international levels. I served in a number of positions at UNESCO, including chairman of the board of a special committee examining the future of the organization and the Brazilian delegate to UNESCO's executive board. It was during this time that I became more familiar with TWAS and got to know Salam on a more personal basis.

A promise from UNESCO

In the late 1980s and early 1990s, TWAS experienced two major events that involved UNESCO. First, there were Salam's efforts to become director-general of UNESCO, which ultimately proved unsuccessful. Salam, who belonged to a minority Islamic sect, failed to receive the endorsement of his home country of Pakistan. Without his government's backing, his candidacy was doomed.

During this period, I invited Salam to speak to UNESCO's executive board. This small gesture marked an important step in strengthening my relationship with him. By laying the groundwork for greater interaction in the future, the event proved valuable several years later when Salam endorsed me to succeed him. His decision was based in part on his personal knowledge of me acquired largely through our involvement with international organizations, especially UNESCO.

While I staunchly supported Salam's candidacy and was disappointed when his effort fell short, I was delighted that the person who was ultimately elected UNESCO's director-general, Federico Mayor, proved to be a staunch supporter of TWAS.

Mayor, a Spanish-born professor of molecular biology who had previously served as UNESCO's assistant director-general, was a strong proponent of science within the organization. He valued the contributions that the Academy was making to the global science community – he was elected a TWAS Fellow in 1991 – and was keen to see the Academy succeed. As a result, when TWAS faced another major crisis in the early 1990s – the IAEA's decision to no longer serve as the lead UN agency providing administrative support for ICTP and TWAS – Mayor assured a worried Salam that UNESCO would stand behind both organizations and assume the responsibilities that IAEA was relinquishing.

In 1990, I completed my term as the head of UNESCO's executive committee and returned home to become a research leader in the Brazilian Centre for Physics Research (CBPF). Once more, I was ensconced in the research community and engaged in intriguing studies, which I never had time to pursue while fulfilling my administrative responsibilities with international organizations.

Brazil was again shaken by a series of dramatic political events. In 1992, the president of Brazil was impeached under charges of corruption. Itamar Franco, the republic's vice-president who hailed from Minas Gerais, replaced him. Amidst the turmoil and uncertainty, I was appointed minister of science and technology, a post I held from 1992 to 1998.

Following Salam

I was elected a member of TWAS in 1988 at a time when I began to take a closer look at the Academy's activities, most notably through the lens of UNESCO.

In the early 1990s, Salam grew increasingly ill as he fell victim to a chronic and progressively debilitating Parkinson-like disease. Under Salam's guidance, I was first selected vice-president of the Academy in 1992. By the time of the Academy's 10th anniversary celebration in 1993, it had become painfully clear that Salam could no longer oversee the Academy's day-to-day activities. Mohamed H.A. Hassan, a Sudanese-born mathematician who had served as the Academy's executive director since its inception, took over the reins of the organization.

Salam, of course, deserves full credit for conceiving the idea of an academy for scientists from developing countries and then carrying that idea forward to levels of achievements that seemed highly improbable at the time of TWAS's creation. However, the place that TWAS now holds in the international science community as one of the world's preeminent institutions dedicated to scientific capacity building in the South is due in equal measure to the talents and efforts of Hassan. He took the mantle from Salam and brought the Academy to new heights. Salam and Hassan deserve equal recognition as the organization's primary architects.

In 1996, I was officially elected president. Salam, despite his illness, played a key role in the process. I believe that he chose me as a replacement largely because of my administrative experience in international organizations (notably UNESCO) and my broad array of contacts in the global scientific community. He was also keenly aware that TWAS's presence was much stronger in Africa and Asia than it was in the Americas. He thought that I could help bring greater balance to TWAS's reputation. Such a recalibration, he believed, would benefit the Academy and, more generally, science, especially in terms of South-South cooperation.

I was both delighted and honoured to be asked to head TWAS, although I realized that I would never be able to replicate Salam's global presence or match his extraordinary level of success. Indeed I was worried that, in trying to follow in his footsteps, I would stumble badly.

Despite my personal concerns, I nevertheless hoped that I would be able to move the Academy ahead in ways that would continue to advance Salam's dreams for a more peaceful and prosperous world – attained, in large measure, by strengthening science in the developing world. In retrospect, I believe that one of the main accomplishments of my presidency was to facilitate the life of TWAS by keeping the Academy relevant and vibrant at a time when science in the developing world was advancing at an unprecedented pace. I viewed what I was I doing as largely a continuation of the job that Salam had begun. In this sense, I became part of the TWAS story, inserted into a script that had been largely written by a brilliant author who had developed the major themes and plot lines well before my arrival.

TWAS faced several critical issues during my presidency. There was the initial challenge of navigating a smooth transition to the post-Salam years and assuring Academy supporters that the organization would continue its good work despite the retirement of its illustrious leader. Salam could not be replaced. Yet the Academy had to prove that it could endure without him. That is why I largely viewed my role as continuing Salam's legacy and convincing others that the Academy could effectively carry on his work.

Another critical issue was funding. This was a persistent challenge throughout the Academy's early decades and it remains a key challenge today. In the early 1990s, the Italian government faced a severe financial crisis that raised doubts about whether it would continue to fund ICTP and TWAS. While the Italian government eventually decided to do so, the Academy recognized that it would remain particularly vulnerable to potential budgetary crises in Italy because its funding was categorized as temporary and therefore subject to annual reviews that often intersected with yearly shortfalls in overall government revenues.

TWAS took two important steps to address this issue. First, it launched an endowment fund that would be built exclusively with contributions from developing countries. The launch of the endowment fund, with an initial target of USD10 million, took place on the occasion of the Academy's 10th anniversary in 1993. Second, the Academy decided to intensify its efforts to transform the Italian government's funding from a voluntary to a permanent contribution. Such a measure would place TWAS's core budget on a much more stable foundation.

I was not involved in either the creation or the fulfilment of either initiative. Negotiations with the Italian government began before my arrival and would continue after my departure. Meanwhile, the endowment fund, which now stands at just over USD12 million, remains an ongoing effort.

Nevertheless, I think TWAS made significant progress meeting both challenges during my presidency. We re-launched the endowment fund campaign just as the initial enthusiasm for the fund began to wane. We made special appeals for contributions from Latin America and, more specifically Brazil, where I had close ties to President Fernando Henrique Cardoso, a lifelong friend and colleague to whom I directly reported as minister of science and technology.

In my mind, raising the profile of TWAS in Latin America and reinvigorating the endowment fund were intertwined issues that I could best address by drawing on one of the critical strengths that I brought to the Academy's presidency: my ties to Latin America's policy community at the highest levels. By showcasing its value to the region, I believed TWAS stood a good chance of obtaining additional contributions to the endowment fund. By the time I left the presidency, the fund had doubled from USD2.5 million to USD5 million.

The same desire to shore up the Academy's funding base and to place it on sustainable path led me to spend a good deal of time discussing with Italian officials how the Academy's funding might become a permanent line item in the government's budget and thus subject to periodic, instead of annual, reviews. Here again I drew on my experience of dealing with government officials during my appointments with UNESCO. While the decision was not finalized during my tenure, I think a great deal of progress was made, laying the groundwork for a change in law that made the Academy's funding permanent. The Italian parliament passed the law in January 2004.

My frequent discussions with Italian officials had another beneficial consequence. In 1999, UNESCO had tentatively named Mohamed Hassan the next director of its regional office in Cairo, Egypt. The appointment meant that he would be leaving TWAS. As president of the Academy, I had experienced first-hand the irreplaceable role that Mohamed had played in the Academy's success and firmly believed that TWAS could not afford to lose him. I made a plea to Italian officials and also contacted UNESCO's Director General Mayor to assert how devastating the decision would be for TWAS. The tentative decision was reversed, allowing Hassan to remain as the Academy's executive director. As a result, Hassan would continue to direct the Academy during its greatest period of growth in the first decade of the 21st century.

Bigger presence, bigger impact

While president of TWAS, I also devoted considerable time to enhancing the Academy's presence in the international and economic development communities. My goal was to broaden the focus of TWAS from issues related to scientific capacity building to issues related to science policy. I shared a belief, held by many others in the Academy, that TWAS had reached a stage of development where it could successfully engage national governments and international organizations in broad discussions of science-based sustainable development. In so doing, TWAS could magnify both its presence and impact.

Our thinking was that the Academy's members not only had a great deal of scientific knowledge to share with others, but that they also had a great deal of experience in successfully navigating the science policy landscape in the developing world. Due to its credibility and track record, and thanks in large measure to the size and quality of its membership, we firmly believed that the Academy, in fact, could serve as an important bridge between the world of science and the world of policy. TWAS was poised, as my successors C.N.R. Rao and Jacob Palis have aptly described, to become the "voice of science" for the South.

That is why during my tenure I focused attention on strengthening the Third World Network of Scientific Organizations (TWNSO). Another creation of Salam's, TWNSO was launched in 1988 to serve as the political and diplomatic arm of the Academy.

TWNSO consisted of research councils, science academies and, most significantly, ministries of science and technology. To bolster the organization, we expanded TWNSO's membership to some 150 institutions. We also launched a major project with the United Nations Development Programme's (UNDP) Special Unit for South-South Cooperation to highlight best practices in applications of science and technology to address critical social problems; organized special ministerial sessions at TWAS's general conferences that provided opportunities for public officials and scientists to exchange ideas; and sought to devise collaborative activities with UNESCO, UNDP, the Global Environmental Facility (GEF) and other international organizations as part of a larger effort to raise the profile of both science and TWAS in economic and finance communities in the developing world.

In all of these efforts, TWAS and TWNSO operated as Salam had envisioned: as complementary organizations that shared the same goals and, in most instances, the same resources and staffing. The synergies between TWAS and TWNSO helped both organizations to advance their overlapping agendas.

In a similar vein, the Academy took initial steps to increase the number of social scientists who were members. While social scientists were always eligible to be elected into the Academy, TWAS had only a few members who were not natural scientists. Among this select group was Fernando Henrique Cardoso, the president of Brazil. Trained as a sociologist, Cardoso was elected to TWAS in 1984. But Cardoso was an exception. In fact, at the time, he was only one of two social scientists who were members. (Arthur Lewis, a Nobel laureate in economics from the United States, was the other.) Given the Academy's roots and mandate, natural scientists, not surprisingly, dominated the organization.

During my tenure as president, we began to devise a strategy – including identifying preeminent social scientists in the developing world – that was designed to raise the number of social scientists in TWAS. The effort would gain momentum after my departure, but the seeds for this initiative were planted during my tenure as president.

In many ways, this task has proven as challenging as TWAS's initial efforts in the 1980s to identify natural scientists in the developing world suitable for election into the Academy. But I believe it must be diligently pursued if TWAS's twin goals of becoming more inter-disciplinary and socially relevant are to be achieved.

With the financial support of the Brazilian government, in 1997 the Academy also launched the Celso Furtado Prize in Political Economy, which is given to scholars who have made critical contributions to understanding and promoting socio-economic development in developing countries. The award is named after one of Latin America's most renowned economists. Furtado earned international acclaim for his studies of the causes of underdevelopment and poverty in developing countries and for outlining the steps that needed to be taken, largely through government action, to overcome these problems.

All of these efforts were intended to increase TWAS's global presence by forging close collaborations with individuals and organizations that both strengthened and complemented the Academy's existing strengths.

My experience with international organizations and, even more so, in science policy circles in Brazil, has shown me that scientific capacity building for the sake of science is a useful, yet insufficient, measure if science is to become a key instrument for progress. For science to attain its highest value within society, proponents have to illustrate how their work will positively impact the lives of people.

Moving on

In 1999, I was appointed Brazil's ambassador to UNESCO. I believed that by accepting the post I would have to resign my position as president of TWAS. In my mind, UNESCO's oversight of the Academy would have created unresolvable conflicts of interest in fulfilling my dual responsibilities. As a result, I tendered by resignation, which was made effective at the conclusion of TWAS's General Meeting in Iran in 2000. I was delighted that C.N.R. Rao, one of the world's most eminent chemists and an unrivalled advocate for science in the developing world, would replace me.

My four-year tenure as president of TWAS generated untold personal rewards. It was indeed a pleasure to lead such an invaluable and effective organization, and it was truly a source of personal pride to succeed such an eminent figure as Abdus Salam. I can only hope that I lived up to the expectations that he had established for the Academy and that I was able to carry forth Salam's dream in ways that he would have appreciated and applauded.

It was a joy to work with the TWAS members, who are the core source of the Academy's strength and success. I would like to express special gratitude to Mohamed Hassan, whose talents and dedication are largely responsible for making the Academy what it is today: one of the world's most respected institutions for the promotion of science and science-based development in the South. I am delighted to have played a role – however modest – in this effort, and I have watched, with both satisfaction and fascination, as TWAS's presence has grown even more prominent since my departure.

Science is not the only factor that determines the well-being of a society. But, as the Academy has illustrated throughout its history, science is certainly one of the key factors in the quest for a more equitable and prosperous world. This, as much its membership, awards, prizes and prestige, is part of the Academy's legacy. And this, as much as any other factor, will serve the Academy well in the years ahead.

TWAS was conceived as a scintillating idea. That idea, I am convinced, will continue to light the way in the years and decades ahead.

ICTP
International
for Theoretical

A centre for excellence

C.N.R. Rao

C.N.R. Rao is one of the world's leading chemists and materials scientists. He has been a prominent advocate for science in the South and a key figure in TWAS's success. He is a Founding Fellow of TWAS and served as president of the Academy from 2000 to 2006.

Born in Bangalore, India, in 1934, Rao received a Bachelor of Science degree from the University of Mysore in 1951 and a Master of Science degree from Banaras Hindu University in 1953. He then travelled to the United States, earning a doctorate degree in chemistry from Purdue University in 1958.

After completing a year of postgraduate study at the University of California at Berkeley, Rao returned to India in 1959 to assume a research position with the Indian Institute of Science (IISc) in Bangalore; he served as IISc director from 1984 to 1994. In 1989, he became founding president of the Jawaharlal Nehru Centre for Advanced Scientific Research (JNCASR), a position he held for more than a decade. He is currently JNCASR's national research professor, Linus Pauling professor and honorary president.

Rao is recognized in the international scientific community for his path-breaking work on the characterization, synthesis and design of new materials. He has been a prominent figure in shaping the contours of materials science, lending his knowledge and skills to a frontier scientific field that has had a significant impact on the global economy. His life-long commitment to strengthening scientific research and education in India has made him an influential figure in his home country, where he has held numerous posts, including chair of the Prime Minister's Science Council.

In recognition of his scientific accomplishments, Rao was elected to the Royal Society in the UK in 1982. Over the past three decades, he has earned membership in more than 25 science academies, including all three Indian science academies (where he has been president of each), the Brazilian Academy of Sciences, Chinese Academy of Sciences, French Academy of Sciences, Japan Academy, Pontifical Academy of Sciences and US National Academy of Sciences. He has also been president of the International Union of Pure and Applied Chemistry (IUPAC) and a member of the executive board of the International Council for Science (ICSU).

He has received more than 60 honorary degrees and has won numerous prizes for his contributions to science. Among his most prestigious awards are the Bharat Ratna (Jewel of India), Indian Science Prize, and SN Bose Medal in India; the Khwarizmi International Award in Iran; the Dan David Prize in Israel; the Ernesto Illy Trieste Science Prize in Italy; the Nikkei Asia Prize for Science in Japan; the Hughes and Royal Medals of the Royal Society and the University of Cambridge's Distinguished Research Professorship in the UK; and the United Nations Educational, Scientific and Cultural Organization's (UNESCO) Einstein Gold Medal.

Rao has published more than 1,500 research articles and 45 books, including such seminal works in his field as Phase Transition in Solids, New Directions in Solid State Chemistry, Chemical Approaches to the Synthesis of Inorganic Materials *and* Transition Metal Oxides. *He has also written several children's books designed to boost interest in science among young people in India.*

In the following essay, Rao traces the events that led to his involvement with TWAS. He also talks about accomplishments during his tenure as TWAS president. He concludes with a broad assessment of the challenges that the Academy is likely to face in the years ahead due the dramatic changes that have taken place in science in the developing world over the past 30 years.

***

I first met Abdus Salam in the late 1960s at a symposium held at the California Institute of Technology in the United States. We spoke at the same session and had an opportunity to exchange pleasantries afterward. Salam, who was then in his 40s, was a well-known figure in science. But he had yet to win the Nobel Prize, which would make him an international celebrity in the worlds of science and diplomacy. That honour would take place in 1979.

I was in my early 30s and just beginning my career after having earned a doctorate degree in chemistry at Purdue University in the United States. Following a one-year postdoctorate appointment at the University of California at Berkeley, I returned to India in 1959 to assume a faculty position with the Indian Institute of Science (IISc) in Bangalore.

While I had received excellent job offers in the United States, I was drawn back home by the allure of an independent India – so forcefully expressed in words by Mahatma Gandhi and in deeds by the nation's first prime minister, Jawaharlal Nehru. I held deep convictions for my native land and a burning desire to be part of its growth and prosperity. It was where I wanted to be.

My vision of the future of science in India focused on two goals – one concrete and one visionary – that I believed to be eminently achievable, indeed necessary, to ensure a better future for my home country: to build a world-class research centre and to make India great in science.

IISc was India's foremost research institution. But neither the city of Bangalore nor IISc had yet gained an international reputation for research and innovation. In the late 1960s, both were largely backwaters in international science. Their star-turn would not come for some time. Science in India held great potential, but fulfilling that potential would require a great deal of work.

The way things were

My first encounter with Salam gave few hints of what was to follow. In fact, it was not until more than a decade later that our paths would cross again.

In the early 1980s, I had learned from colleagues in India that some of the developing world's most accomplished scientists, led by Salam, had met at the Pontifical Academy of Sciences to discuss the prospects for organizing an academy for scientists from the developing world. I soon received an unsolicited letter from Salam congratulating me for being selected a Founding Fellow of the Third World Academy of Sciences (TWAS). I was one of 42 scientists chosen.

The selection process for membership to TWAS, in the years immediately following the Academy's creation, was quite informal. In fact, TWAS's initial class of fellows was comprised of scientists who had already achieved broad recognition for their work by virtue of having been elected members of the world's most prestigious science academies, most notably the Royal Society in the UK, the National Academy of Sciences in the US and the Pontifical Academy of Sciences in the Vatican. All of these academies, not surprisingly, were located in the developed world.

Election to a prestigious science academy in the North was undoubtedly the surest path to recognition for scientific achievement in the South – a state of affairs that was just one more reflection of the lowly state of science in the developing world.

Based on the criteria that TWAS had devised, my election to the Royal Society in 1982 had made me a candidate for membership to TWAS. I was from the South, which rendered me eligible for membership to the Academy. Nevertheless it was the honours that I had received in the North that gave me the scientific credentials warranting my selection.

From a practical perspective, the thinking was that TWAS's members had to be selected quickly if the Academy was to get off the ground and running in rapid fashion. In fact, a deliberate review process was impossible in the Academy's inaugural year. TWAS simply had no members on hand to elect new members.

A methodical selection process, moreover, remained extremely difficult for the first several years when TWAS members numbered less than one hundred. The absence of a critical mass of scientists in the Academy – and extremely low representation in a number of fields – dictated how to proceed.

What was true of TWAS membership also applied to the Academy's grants and prizes programmes. These programmes were launched in the mid-1980s, following the Italian government's decision to provide funds for TWAS's administration and activities.

Rules and procedures for the nomination and selection of potential candidates did not exist. Awareness of the Academy's programmes was not widespread, and getting the word out about TWAS's grants and prizes programmes was a difficult and time-consuming task. As a result, the likelihood was that few scientists would apply on their own volition. TWAS would have to be proactive in identifying worthy nominees.

I am not suggesting that such an informal way of doing business represented a weakness of TWAS during its early years. It was simply a reflection of the way things were. TWAS, in fact, was a highly principled organization driven by magnificent ideals and high standards. But the Academy could not avoid the facts on the ground if it wanted to succeed.

Serving excellence

As mentioned earlier, being elected a member of a prestigious science academy in the North provided an excellent proxy for identifying worthy candidates for membership to TWAS. Similarly, the opinion of well-respected scientists, including those who had been elected to TWAS, was an effective way to decide which scientists should be considered for a TWAS grant or prize.

This word-of-mouth informality, in many ways, served the Academy well during its early years. It allowed TWAS to establish its credibility in short order without compromising the quality of its members, grantees and prizewinners. The process may have been informal, but it was not casual. In fact, it was rigorous in its own way.

Looking back, I think it is only fitting that the Academy's reputation from its earliest days depended largely on the reputation of Salam, and that Salam himself chose to place his reputation on the line in the creation and early development of TWAS. If ever an institution was due to the vision and determination of one individual, it is TWAS.

However, Salam knew that word-of-mouth and collegial exchanges of information and opinions could go only so far over the long run, and that the Academy's success could not be sustained without constructing a formal framework for its operation.

That is why he sought to create official statutes and bylaws for TWAS soon after the Italian government announced in 1985 that it was giving the Academy a three-year, USD3 million grant to support its operations and activities. With TWAS's immediate future no longer in doubt, it was time to build a more formal and transparent framework for its administration.

By then, I had become more actively involved in the Academy. I had attended TWAS's founding meeting in 1983 at the University of Trieste and then participated in TWAS's first international conference and official launch in 1985, held on the campus of the International Centre for Theoretical Physics (ICTP), another remarkable organization that Salam had created.

A highlight of the conference was the opening address by then United Nations Secretary-General Javier Pérez de Cuéllar. His presence – as well as the presence of other dignitaries – not only reflected Salam's elevated status in the international scientific community, but also indicated that the Academy had struck a responsive chord among those deeply involved in issues related to science and development in the South. Salam not only drew eminent scientists into the Academy's fold. From the start, he also succeeded in convincing outstanding institutional partners to stand shoulder-to-shoulder with TWAS in their pursuit of shared goals.

During these early years, I sought to identify scientists, especially those in my field, whom I viewed as excellent candidates for membership in the Academy. In addition, I had participated in several small workshops that TWAS had organized on materials sciences and I had lent a hand in editing the workshop proceedings.

Held in a variety of subject areas, the workshops were led by eminent scientists in their fields. In attendance were also a select group of young, promising researchers from the developing world. There usually were no more than 10 to 15 participants.

These workshops were an early example of how the Academy could bring together scientists from around the world to exchange ideas in a collegial atmosphere where frontier research questions were scrupulously examined. It was a wonderful experience, one that I have never forgotten.

Learning from Salam

The most important aspect of my involvement in the early activities of TWAS, however, was that it allowed me to speak with Salam, both frequently and at length, about the future of the organization and, more generally, the prospects for building scientific capacity in the developing world.

Despite Salam's grand ambitions, he was determined to present TWAS – and the circumstances that the Academy faced – in the clearest terms possible. He did not want others to question or misinterpret the core purpose and central focus of the Academy.

As a result, he insisted that we forego such obscure terms as 'the South' and instead simply call the organization the 'Third World Academy of Sciences'. For Salam, this was to be an academy of and for scientists from poor countries. At the time of TWAS's creation, Salam did not view 'Third World' as a derogatory term. In fact, I am not sure that he ever did. For him, it was a statement of fact.

At the same time, Salam was particularly keen that this bold statement of fact did not devalue the Academy's most valuable currency: its uncompromising commitment to scientific excellence in poor countries.

From the Academy's inception, Salam was saying to the international scientific community, in effect: "We are who we are, but we can be your equals. What we lack are the opportunities and resources to nurture and apply our talents."

This strategy not only provided a way of recognizing the difficult realities that the Academy faced, but it also instilled a sense of pride and confidence among scientists living and working in developing countries.

Salam exhibited the same candour and courage when attending the Nobel Prize ceremony in Stockholm in 1979. He came to the ceremony attired in traditional Pakistani dress, including a turban. He also brought his two wives, ensuring that his entire family would share in the joy of the event.

The attendance of his two wives placed the organizers of the Nobel Prize ceremony in an uncomfortable position. They had no desire to insult one of their prizewinners by denying his wishes; nor did they want to challenge the conventional norms of their guests at a gala event designed to celebrate the highest achievements of science.

In the end, the presence of Salam's two wives at the ceremony was a deep reflection of his cultural roots and a compelling way to remind the world of the universality of science. "I am a scientist," Salam seemed to be saying, "yet I am also a citizen of Pakistan and a devout believer in the Islamic faith."

The irony is that while Salam was unabashedly celebrating his own culture and religion in science's most revered setting, he was being ostracized in his home country for his faithful allegiance to the Ahmadiyya sect, which many of his fellow Muslims believed resided outside the true pillars of their faith.

Immersed in the Academy

Salam was never one to ignore enthusiasm and commitment. Since I had displayed both during the earliest days of the Academy, he decided to ask me to head the committee responsible for drafting TWAS's statutes and bylaws.

"Ask", if truth be told, is too gentle a way to describe his request. As was often the case with Salam, when he wanted something done, he simply and persuasively commandeered your services. He would take you aside and quietly beseech you to *please* do something for the sake of the Academy.

Yet, what he really meant was that you *must* do something for the Academy – that it was an obligation you needed to fulfil as part of your commitment to science and society. Once you agreed, he would quickly and methodically follow this "request" with a description of what needed to be done – without delay.

Compared to my previous activities, the drafting of the statutes and bylaws marked my first extensive involvement with the Academy, one that immersed me in the fundamental principles and workings of TWAS. Not surprisingly, it exerted an effect on how I have viewed and interacted with the Academy ever since.

This also marked the first time that I had an opportunity to work closely with Mohamed H.A. Hassan, who would go on to serve TWAS with such distinction for nearly three decades, at the outset as executive secretary and then as executive director.

Salam had initially invited Hassan, a young mathematician from Sudan, to Trieste in the mid-1970s as an ICTP Associate. Hassan became involved with TWAS soon after the meeting at the Pontifical Academy of Sciences, marking the Academy's launch. He was made a permanent staff member in 1986 following the Italian government's pledge to provide financial support for the Academy. I remember sitting in on the meeting where Salam officially interviewed Mohamed for the post. Salam pulled me aside afterward to say that he thought we should take him, something that we all knew he should and would do.

The statutes and bylaws, which were drafted in 1986 and 1987, marked an important step forward for the Academy. As in all organizations, such documents helped to regularize the procedures and methods of operation.

In TWAS's case, the statutes and bylaws outlined the goals and purpose of the Academy, laid out the selection procedures and responsibilities of the various officers, described the primary activities of TWAS and crafted a framework for the election the members.

These were all complex challenges. However, I remember one issue, in particular, that proved particularly vexing.

A question of membership

In the informal atmosphere that characterized much of TWAS's decision-making during its early years, three different categories of membership had been established: 'Fellows' who were from the South; 'Associate Fellows' who were from the North; and 'Corresponding Fellows' who were either young scientists with promise or more mature scientists with respectable accomplishments that were nevertheless deemed insufficient to warrant election to TWAS as full-fledged members.

The truth is that the criteria separating Fellows from Corresponding Fellows were not precise. Formal measures for distinguishing one category from the other had not been clearly articulated in either discussion or in writing.

Since a Corresponding Fellow could be viewed as a Fellow-in-waiting, this category of membership carried risks for the Academy. It could elicit criticism, especially among well-regarded scientists, for its implicit ambiguity and slight.

And that is exactly what happened when the president of the Indian Academy of Sciences was selected as a Corresponding Fellow. Interpreting this honour as a snub, he declined the invitation.

I discussed this issue with Salam, contending that Corresponding Fellows should not be included as a category of membership because of the vagueness of the criteria and the problems – and embarrassments – it might cause.

Salam agreed. As a result, the Academy adopted just two categories of membership: Fellows and Associate Fellows. Designations of Corresponding Fellow would disappear in the early 1990s. The distinction between Fellows, who were from the developing world, and Associate Fellows, from the developed world, was ended in 2013.

The difficult issue of determining who was worthy of election to TWAS was ultimately resolved by the creation of membership advisory committees in broad fields of research ranging from the agricultural sciences to physics. The committees were assigned the task of assembling and rigorously evaluating an annual slate of candidates and then submitting their recommendations to the TWAS Council for final approval. These committees, which became fully operational in 1992 prior to the Academy's Fourth General Conference in Kuwait, have been a cornerstone of the election process ever since.

Members electing members, through an uncompromising process of evaluation based on excellence, has been the hallmark of merit-based scientific academies around the world for centuries. Indeed it is the defining characteristic of these learned institutions. The strategy has served TWAS well, ensuring that the Academy is comprised of the developing world's most eminent scientists.

The high standards that have infused the Academy's membership election process have been largely responsible for the sterling reputation of TWAS. It has set the stage for the unimpeachable quality of the Academy's membership, which, in turn, has created a strong and credible platform for the creation of TWAS's impressive array of programmes for research grants, fellowships and awards. TWAS's renowned members have also enabled the Academy to serve as a trustworthy and authoritative voice on a broad range of issues related to science and development.

The breadth and reach of TWAS originates with the quality of its membership. Both the informal procedures for membership that marked the selection process during the first few years, and the more formal election process that has characterized the Academy ever since, have been predicated on this principle.

The end of an era

Salam was able to accomplish virtually all that he did with TWAS in a less than a decade. By the early 1990s, he was showing signs of distress caused by a degenerative Parkinson-like malady that first attacked his mobility and later his speech.

Nevertheless, throughout this period and until nearly the very end of his life, Salam remained intensely involved in the Academy's affairs. He led TWAS during a challenging time in the early 1990s when administrative responsibility for the Academy was transferred to the United Nations Educational, Scientific and Cultural Organization (UNESCO) and when funding was temporarily halted by a budget crisis in Italy that called the Academy's future into question. He also presided over the Academy's general meetings in Venezuela in 1990 and Kuwait in 1992. In addition, he led the activities at TWAS's 10th anniversary meeting in Trieste in 1993, where the Academy's endowment was officially launched.

In 1991, Salam was at the centre of efforts to secure a grant from the Swedish International Development Cooperation Agency (Sida) to support the research of young scientists from the South. This proved to be the first step in an ongoing partnership with Sida that has continued to this day. Only the Italian government's contributions to TWAS now exceed those of Sida's. In 1993, Salam was also at the centre of discussions for the creation of the TWAS-Indian Council for Scientific and Industrial Research (CSIR) fellowship programme – the first of TWAS's South-South postgraduate and postdoctoral fellowship programmes. The initiative has since become one of the cornerstones of the Academy's capacity-building efforts.

As Salam eventually became too ill to lead TWAS, Mohamed Hassan, the Academy's executive director, began to shoulder increasing responsibilities. Thanks to Mohamed's talents and abilities, TWAS continued to execute its programmes in exemplary fashion. It also continued to actively participate in international scientific meetings, including the Rio Earth Summit in Brazil in 1992 and the launch of the Commission on Science and Technology for Sustainable Development in the South (COMSATS) in Pakistan in 1994.

Nevertheless, as Salam grew increasingly ill, the TWAS Council began to contemplate plans for having someone succeed him. The procedures for succession were not self-evident. The Academy, as a result, again turned to an informal decision-making process, and such informality would again serve the Academy well.

Some might attribute the outcome to good fortune. But I believe that it was a reflection of the quality of the people who were involved and the level of thought and consideration that went into the decision.

Salam himself was engaged in this process, which must have been extremely painful for him to do. After much discussion and negotiation, in January 1995, José I. Vargas, the Minister of Science and Technology, was named interim president and, in September 1995, he presided over the TWAS General Meeting in Nigeria.

Vargas's extensive diplomatic skills and expansive network of international contacts would prove a boon to the Academy in the years ahead. It is often said that the most important decision a leader makes is choosing a successor. Salam, to his credit, chose well.

I last met Salam in early 1996 when visiting Trieste. He was extremely ill at the time. Communication was difficult. But he did manage to convey his thoughts, sometimes speaking in Hindustani (Urdu), which I also speak. I remember him motioning me to come closer and to dip my shoulder so I could hear his words more clearly. He clenched my hand and in a whispering, halting tone, said: "Please take care of the Academy."

Within months Salam would be dead. News of his passing was announced just days before TWAS's Eighth General Meeting, in Trieste. The event was turned into a homage for Salam with tributes delivered from the around world.

Salam had left behind an enormous legacy both as a scientist and humanitarian. He had attained the highest echelons of respect and honour within the scientific community as a Nobel laureate. He had been the driving force behind the creation of ICTP, an institution that helped to redefine what it meant to be a global centre for scientific research and training. And, of course, he had conceived and then instigated the establishment of TWAS, which, under his tutelage, had emerged as a unique and powerful institution for the promotion of scientific capacity building and excellence in the developing world.

Any one of these achievements would have been reason to speak in praise of an extraordinary life. All three defined a singular, unmatched life of accomplishment.

Transitions and progress

With the death of its iconic leader, the future of the Academy, despite all of its good fortune, had been placed in doubt. By virtue of his towering reputation and his boundless contributions, Salam's name had become synonymous with TWAS. It was not frivolous, then, to ask whether the Academy could survive – let alone prosper – in his absence.

José I. Vargas proved to be more than up to the challenge. With the help of the TWAS Council and Hassan, he ensured that TWAS remained on solid footing during this critical transition in leadership.

During this period, I served as the Academy's vice-president, working closely with Vargas. I also continued to oversee the Academy's elections committees. Vargas gave me full rein in executing my responsibilities, helping to ensure that the Academy's emphasis on excellence would continue unabated.

In May 2000, after being appointed Brazil's ambassador to UNESCO, Vargas unexpectedly announced that he would step aside as TWAS president in October 2000. He reasoned that simultaneously holding the positions of TWAS president and UNESCO ambassador posed a potential conflict of interest. He worried that others would contend that, in fulfilling his oversight responsibilities of TWAS at UNESCO, he could not remain impartial when addressing issues related to the Academy. That sentiment, he cautioned, would always hover in the background regardless of how fair-minded he might be.

As vice-president of TWAS, I was in the direct line of succession for the presidency and, with Vargas's departure, I became the Academy's third president. It was both an honour and a privilege to assume this responsibility.

As president, Vargas had made progress on a number of critical issues facing the Academy. Therefore one of the major tasks during the first years of my presidency was to continue to push ahead on many fronts and to bring many of the negotiations that Vargas had begun to a successful conclusion.

The years 2003 and 2004 may have been the most momentous in TWAS's history since the Academy's inaugural year in 1983 and its official launch in 1985. A whirlwind of events and decisions solidified the Academy's financial base, broadened the scope of the Academy's activities, and raised TWAS's profile within the scientific community to un precedented heights.

TWAS at 20

In October 2003, the Academy held its 20th anniversary meeting in China. The opening ceremony, which took place in the Great Hall of the People, was highlighted by a speech given by Hu Jintao, president of China. More than 3,000 people were in attendance. In many respects, this event signalled the arrival of TWAS as a key player in international science. At the same time, the successful launch of the first Chinese astronaut (*taikonaut*), which took place on the opening day of the conference, further confirmed the arrival of China's scientific community on the world stage.

The growth of TWAS and the growth of science in the developing world had parallelled one another over the course the past two decades. By the dint of coincidence, these two trends intersected for all the world to see at the Academy's 20th anniversary meeting. Salam would have been pleased.

Just three months later, the Italian government announced that its annual funding allocation to TWAS had been altered from a voluntary to permanent contribution. This decision ensured that the Academy would have a strong and stable financial foundation from which to grow. The announcement, published in Italy's national register of governmental laws and regulations in January 2004, represented the successful culmination of lengthy discussions with the Italian government that had begun in earnest during Vargas's presidency and culminated during mine. No other decision during my tenure as president would be as crucial to the Academy's long-term financial health and vitality.

Between 2003 and 2005, TWAS also launched five regional offices in Brazil, China, Egypt, India and Kenya. Each office was housed at a major scientific institution within the country, providing an important focal point for scientific collaboration between the Academy's secretariat and scientific institutions throughout the developing world. For example, the TWAS regional office for Central and South Asia is headquartered at the Jawaharlal Nehru Centre for Advanced Scientific Research (JNCASR) in Bangalore.

The regional offices have played an increasing role in the Academy's efforts to decentralize its activities. I believe that this trend will remain essential if TWAS hopes to meet the growing demands for scientific capacity building and South-South cooperation in science in the years ahead.

Decentralization will be particularly important in the administration of the TWAS's fellowship grants for PhD students and postdoctoral scholars. This core activity of the Academy took a great leap forward in 2003 when Brazil, China and India announced that they would each provide 50 grants annually for students from other developing countries to attend universities and research centres within their own country for training. The host countries cover the cost of tuition and hospitality. TWAS covers the students' transportation costs. Many other countries – for example, Malaysia, Mexico, Pakistan and South Africa – have since joined the programme.

With more than 500 fellowships currently offered each year, the TWAS fellowship programmes have become one of the largest initiatives of its kind for South-South cooperation in science. As the programme continues to grow, much of the management for its operation will likely shift (indeed, by necessity, will have to shift) to the Academy's regional offices. This devolution of responsibility will take place both for practical reasons and as a reflection of the growing strength of all facets of science in the developing world, including the administration and management of scientific activities.

Expanding context

Over the course of my early years as president, other events – both well within the Academy's reach and far beyond its ability to control or influence – also impacted TWAS.

One week after I became president, it was announced that the InterAcademy Panel (IAP) would be coming to Trieste from its interim home at the Royal Society in the UK and that its administrative operations would be placed under the umbrella of TWAS. This decision not only increased the number of institutions partnering directly with Academy, but it also helped to broaden the reach of the Academy from the South to the North

– in the process strengthening TWAS's position as a significant player in international science. In 2004, the InterAcademy Medical Panel (IAMP), carrying a mandate similar to IAP but focusing on the medical sciences, would likewise bring its secretariat to Trieste. IAMP would also be placed under the administrative umbrella of TWAS, marking yet another step in broadening the Academy's influence.

In 2002, TWAS assumed a prominent role in the World Summit on Sustainable Development (WSSD), working alongside IAP and the International Council for Science (ICSU) in organizing sessions that examined the role of science in development, especially in the developing world – again strengthening the Academy's voice as a knowledgeable advocate for science in the South.

The world was paralysed by the tragic events of 11 September 2001. It was a horrendous moment that struck all of us, some in deeply personal ways. My wife's nephew, who worked in the Twin Towers, lost his life that day. The dreadful event led TWAS to postpone its planned conference in India to 2002.

The Academy not only spoke out against such ghastly acts but also noted that science could help counteract the forces that gave rise to such horrific behaviour by serving both as an instrument for economic growth and as a source of education that could lead to greater understanding and tolerance.

As I wrote in the *TWAS Newsletter* several weeks after the attacks: "I am convinced that whenever and wherever fundamentalism dominates, blind faith clouds objective and rational thinking. When such forces take hold, they create a mindset that allows people to do unusual – indeed sometimes unspeakable – things.

"That's where science comes into play. My point is that fundamentalism separates science from society and that the big loser in this dynamic is always the society in which the separation takes place.

"It is my belief that the absence of science – and scientific education – leave a void in analytical thinking that is often filled with parochial and antihuman sentiments that can, when mixed with other factors, drive fundamentalists to ram airplanes into buildings."

I reiterated similar sentiments several months later, also in the *TWAS Newsletter*: "The tragic events of 11 September remind us all of how uncertain life is. They also remind us of how important science can be, not just for the material benefits that are derived from scientific research, but for the global community of mutual understanding and respect that is created through efforts to learn more about the natural world and the forces that make it so.

"TWAS has been instrumental in advancing the cause of science throughout the South and it has played an increasingly important role as a bridge between scientific communities in the South and North. As the events of last year show, the Academy's role in the global scientific society has never been more significant."

Strengthening core elements

As TWAS continued to grow and expand, somewhat ironically an increasing portion of my time and energy was focused on strengthening the core elements of the Academy. During my concluding years as president, I continually called on the Academy to "return to basics".

In my mind, this meant reinforcing the Academy's dedication to scientific excellence. This also meant ensuring that TWAS's superb administrative staff would be large enough and skilled enough to meet the growing demands that were being placed upon the Academy. This meant, as well, taking measures to increase the endowment fund so as to provide another sturdy pillar of financial support for the Academy's budget – a source of funding that, for good measure, would be derived from governments in the South. And, finally, this meant seeking outside funds – from foundations and governments both in the South and North – to strengthen the Academy's programmatic activities, most notably its postgraduate and postdoctoral fellowship programmes.

I realized that such activities might not elicit the same level of excitement and enthusiasm that had been generated during the first years of my presidency – a time marked by such historic accomplishments as securing permanent funding from the Italian government, launching the TWAS regional offices and vastly expanding the Academy's postgraduate and postdoctoral fellowship programmes. But I also knew that these activities would be essential to the future well-being of the Academy.

An institution that is fortunate to experience unprecedented growth must also must be wise enough to take the time to consolidate these gains and to fold them into its everyday practices, rendering them a permanent aspect of the institution's activities.

Consequently, I have viewed the efforts that the Academy made to strengthen its administration and programmatic activities between 2004 and 2006 as essential to TWAS's long-term well-being and success. I feel both fortunate and honoured to have played a critical role in leading TWAS through years of profound and lasting change, helping the Academy reach and, equally important, sustain new levels of success and influence.

Pursuing geographic balance

The growing success of the Academy also helped to place the spotlight on challenges that had been largely relegated to the back burner during the Academy's first two decades of existence.

First and foremost among these challenges was this: TWAS's emphasis on excellence – combined with the success of a select number of developing countries (most notably, Brazil, China and India) in building scientific capacity – meant that the Academy had come to be dominated by a few countries.

The developing world has never been a single monolithic block. While all developing countries have shared a disproportionate burden posed by poverty, vast differences in history, culture, religion, governance and population have always been present. Sweeping portraits of the developing world have obscured the South's enormous diversity.

Yet, at the time of TWAS's creation, it was also true that formidable levels of poverty and a startling lack of scientific capacity burdened all developing countries. I think that this is why, in part, Salam was resolute in proclaiming that the organization should be called the Third World Academy of Sciences. The name was a sign of both similarity and solidarity.

That is no longer the case. Over the past two decades, the gap between the scientific capacity of Brazil, China, India and several other developing countries and the 81 countries that TWAS has identified as "scientifically lagging" have become significant. Equally worrisome, this gap, by some measures, has been growing.

It should not be surprising that developments in TWAS have reflected these trends, which have been most clearly manifested by the disproportionate number of Academy members and prizewinners who have hailed from large and increasingly scientifically competent countries.

TWAS's ability to address the growing "South-South" gap in science is no easy matter. The modest level of Academy resources and its small staff have significantly confined TWAS's impact. Meanwhile, TWAS's competing goals of equity and excellence have posed strategic challenges as well.

But this does not mean that TWAS should minimize strategies that help to identify worthy scientists in scientifically lagging countries, or that the Academy should curtail efforts to improve education and training in scientifically lagging countries so that all scientists have an equal opportunity to be elected to TWAS and to receive TWAS prizes.

During my presidency, we tried to broaden the range of participation in TWAS membership and activities in a number of ways without compromising the Academy's devotion to scientific excellence. For example, we intensified efforts to identify scientists working in small, scientifically lagging countries regarded as worthy of election to TWAS, but who had yet to receive the recognition they deserved. We did this by asking Academy members to make special efforts to identify such scientists. We solicited the assistance of our regional offices as well.

The process helped TWAS to elect scientists from countries that had never before been represented in the Academy – for example, Azerbaijan, Bahrain, Malawi and Rwanda. Expanding TWAS's membership to a larger number of countries, I believe, has been good not only for the scientists (and the countries they represent) but also for TWAS.

Despite such progress, significant differences in scientific capacity and excellence among developing countries continue to give distinct advantages to researchers who come from the developing world's most science-proficient countries.

One way to address this challenge is to improve education and training in all developing countries. The TWAS PhD and postdoctoral fellowship programmes have helped to advance this goal. So too has the TWAS

Young Affiliates programme, which has allowed scientists under the age of 40 to participate in Academy activities during their five-year appointments. Since the regional offices have taken the lead in administering the Affiliates programme, the hope is that scientists from small, scientifically lagging countries can be more readily identified because of the offices' close ties with universities and research centres across the region.

Other programmes designed to help young scientists gain the research and training that they need and deserve have included TWAS Prizes to Young Scientists in Developing Countries, the African Union-TWAS Young Scientists in Developing Countries Prize, the TWAS-AAAS Microsoft Award for Young Scientists, the BioVision-Lilly Award given in conjunction with TWAS, and the TWAS-Lenovo Science Prize. There is also the C.N.R. Rao Prize for Scientific Research, which I am delighted to sponsor, and the Atta-ur-Rahman Prize in Chemistry. The latter two prizes are specifically designed to help young scientists working in countries where science has yet to gain adequate support and where conditions for research remain challenging.

I have no doubt that all of these programmes will help to foster greater diversity within the Academy. Yet the full promise of these programmes will take time to be realized. Moreover, other factors – most notably, each countries' willingness to invest in research and promote science education – will likely prove more significant than anything that organizations such as TWAS can do.

Yet I have been convinced, for some time, the Academy must do all that it can to speed the process of scientific capacity building. We cannot sit idly by as the South-South gap in scientific capacity persists and the talents of scientists in poor countries are compromised by limited educational and training opportunities.

That is why I am proud to have led the effort to create the TWAS Research Grants Programme for Least Developed Countries (LDCs), which assists groups of researchers who have proven productive and innovative while working under trying conditions. Specifically, the programme currently provides USD30,000 over 18 months to these groups to help them pursue and expand their research agendas.

The funding may seem modest. Yet it means a great deal to the researchers who are awarded these grants. They are often at critical junctures in their careers, trying to establish a reputation for themselves and the departments and centres in which they work. Of all the programmes that were launched and expanded during my presidency, I believe that this one may ultimately have the greatest impact on helping scientists in science- and technology-lagging countries to compete successfully for recognition and rewards.

As I observed when I stepped down from TWAS in 2006, my involvement with the Academy proved to be among the most meaningful and rewarding experiences in my career. I was delighted to have served as president during such a historic period in the Academy's development and to have concluded my term in office at a time when the Academy was healthier and more vital than it had ever been. I was pleased, as well, to leave TWAS's future in the capable hands of Jacob Palis, the Brazilian mathematician, who led the organization to even greater heights.

Solidarity in a diverse world

The Academy's future seems brighter than it has ever been. TWAS's enviable track record of success and unassailable reputation indicate that it has never been in as strong a position to play a leading role in the international scientific community.

I believe this to be true. Yet I also believe that TWAS would be wise to formulate new ways forward that will allow the Academy to remain as relevant – indeed as irreplaceable – in the future as it has been over the past few decades. I offer this cautionary observation because science in the developing world is changing at an unprecedented pace.

There is no point in having the Academy – in fact, any organization – address bygone challenges, and there is little doubt that the future of science in the developing world will be dramatically different than the past, largely characterized by new issues and new challenges. This is true about all facets of the scientific enterprise, including education, training, research, administration and science-based development and innovation.

As a result, the fact that TWAS has been successful in the past – and continues to be successful today – is no guarantee that it will be successful in the future. It must keep up with the times.

The portion of national gross domestic product (GDP) that China invests in research and development (R&D) more than doubled from 0.09% in 2000 to 1.98% in 2012. Gross expenditures on R&D in China (having exceeded those in Japan in 2006) are now higher than in every other country except the United States. Similarly, Chinese scientists now publish more articles in peer-reviewed scientific journals than colleagues in any other country except the United States. China, moreover, is expected to gain the number-one ranking in publications in a year or two (although the quality of its publications and the number of times the publications are cited by other scientists will continue to lag behind the United States and other scientifically advanced countries for the foreseeable future).

My home country of India has made impressive gains in information and communication technologies and pharmaceuticals. It is now the world's largest producer of vaccines and a world leader in the production of generic drugs. It has embarked on an ambitious programme to expand its research and university systems. India is a youthful country that holds great promise for advancing its scientific capabilities. Brazil has strengths in agricultural science and aeronautics. Its research system has enjoyed a significant boost in investment over the past decade and enrolments in graduate and postgraduate programmes have grown at an impressive rate. Other developing countries and emerging economies – for example, Chile, Mexico, Turkey and South Africa – have also made important strides in building their scientific capacity.

Thus, there is a critical question facing TWAS: What will the developing world's most scientifically advanced countries gain from their continued involvement with the Academy now that they have seemingly passed the tipping point in building their own scientific capacity? Will these countries eventually decide that they can readily achieve their goals for scientific capacity building and science-based development on their own or through bilateral arrangements with other countries? Will they conclude that the need for such brokering agencies as TWAS has become an artefact of history? Will bilateral agreements be embraced as a more effective way to achieve South-South collaboration in science?

Conversely, how can the Academy devise a broad portfolio of effective programming activities given that the developing world is now comprised of a mosaic of nations marked by huge differences in scientific capacity which, by some measures, are as stark as those that exist between the South and North?

Put more succinctly, what will be the role of more advanced scientific countries in TWAS in the future, and what role will TWAS, in turn, have in helping to build scientific capacity and excellence in these countries? In a similar vein, what role will TWAS play in promoting scientific capacity among countries that continue to lag behind in scientific capacity, especially in terms of South-South collaboration? These are not idle, academic questions given the rapid pace of growth in an increasing number of developing countries – a term that itself has become problematic.

Salam faced many seemingly insurmountable obstacles in his noble efforts to promote scientific capacity and excellence in the developing world. Nevertheless, he largely operated in an environment in which there was a great deal of similarity – and solidarity – among developing countries, based on shared experiences and shared conditions. Increasingly, that is no longer the case.

To some degree, TWAS has recognized these trends and has acknowledged that there is no going back.

Such awareness has been reflected in the two name changes that have taken place over the past decade: first, to TWAS, the academy of sciences for the developing world, in 2006, and then, in 2012, to The World Academy of Sciences for the advancement of science in developing countries.

TWAS's new name offers a shorthand way of recognizing this new reality. But it will take more than a name change for TWAS to position itself in the new scientific landscape where North-South divisions in science are rapidly narrowing, at least among some countries, while South-South divisions are becoming increasingly pronounced.

Despite its current success, TWAS's ability to navigate these historic changes in global science will not be easy. But, as Salam would be the first to remind us if he were alive today, neither was building the Academy in the first place.

Fulfilling the promise

Jacob Palis

Jacob Palis is one of the world's most distinguished mathematicians, renowned for his work on chaotic and dynamical systems. Born in Uberaba, Brazil, in 1940, Palis received a bachelor's degree in engineering from the Federal University of Rio de Janeiro in 1962 and then master's and PhD degrees from the University of California at Berkeley, USA, in 1966 and 1967. He returned to Brazil in 1968 after an additional year of post-doctorate study in the United States.

In Brazil, Palis assumed a research position with the Institute for Pure and Applied Mathematics (IMPA) in Rio de Janeiro, becoming a professor in 1971. IMPA has been his home base throughout his career – the place where he has conducted his most important research and supervised more than 40 doctoral students. Serving as director of IMPA between 1993 and 2003, he played a key role in raising the international reputation of the institute.

He has held other key positions in scientific institutions in Brazil as well, serving as a board member of the National Research Council (CNPq) in the late 1980s and again in the mid-2000s and as president (2007-present) of the Brazilian Academy of Sciences.

He has also been involved in numerous international scientific organizations. He has been president (1999-2002) of the International Mathematical Union (IMU); a member of the executive board (1993-1996) and vice-president for science (1996-1999) of the International Council for Science (ICSU); a member of the scientific committee (1990-2006) of ETH (the Swiss Federal Institute of Technology) in Zurich, Switzerland; and a founding member and first chair of the scientific committee (1995) of the Latin American and Caribbean Mathematical Union.

Palis has enjoyed a long-standing presence and held leadership positions among the scientific institutions in Trieste, Italy, most notably the Abdus Salam International Centre for Theoretical Physics (ICTP) and TWAS. He first came to Trieste to help organize an ICTP school on dynamical systems in the late 1970s and continued to direct the school for more than 20 years. He was a member of the Scientific Council at ICTP from 1993 to 2005, including two years as the Council's chair (2003-2005).

Palis became acquainted with TWAS through his association with ICTP. He won the TWAS Prize for Mathematics in 1990 and was elected a member of TWAS in 1991. Over the next decade, he served in various committees for the Academy, including the membership and prize committees. He became secretary-general of TWAS in 2001 and president of TWAS in 2007, leading the Academy until the end of 2012. During his presidency, TWAS's fellowship programmes and endowment fund experienced enormous growth and the Academy became a leading voice for the promotion of science in the South.

He has been the recipient of many honours. He has been awarded the Moinho Santista prize in mathematics and the National Prize for Science and Technology in Brazil; the InterAmerican Prize for Science; the Mexico Prize for Science and Technology; the TWAS Prize in mathematics, the Ernesto Illy Trieste Science Prize, the Accademia dei Lincei Prize in mathematics and the Balzan Prize in mathematics, all in Italy. In addition, he has been elected a member of science academies in Brazil, Chile, China, France, Germany, India, Italy, Mexico, Norway, Portugal, Russia and the United States. He holds honorary degrees from eleven universities and has been named a Chinese Academy of Science Honorary Einstein Professor.

In the following article, Palis traces the steps that led him to TWAS. He highlights the important role that collegiality and partnerships have played in the Academy's success and describes the broad range of activities in which TWAS was engaged during his tenure as president.

***

I 'discovered' Abdus Salam when I was in my late 30s and early 40s. I didn't know who Salam was until I was invited to the International Centre for Theoretical Physics (ICTP) in Trieste in the late 1970s. I am still astonished by the fact that I wasn't acquainted with either Salam or the ICTP until I was well established in my career.

In this sense, my journey to Trieste was quite different from the journey taken by many other scientists of my generation. Most knew of both Salam and ICTP early in their careers. For scientists from the developing world, coming to ICTP was the realization of a dream. As they engaged with others in Trieste who shared that dream, it also confirmed their faith in their own talents and the importance of the work they were doing.

For them, Salam was an iconic figure in science and ICTP was a revered place of learning. To journey to Trieste and to meet Salam in person was an unforgettable experience – a once-in-a-lifetime event and certification that one was a worthy researcher in his or her field of study.

I would subsequently share these sentiments. But when I first came to Trieste, I knew none of that. I only learned about the exalted place of Salam and ICTP in international science circles after my arrival.

Lasting impressions

The invitation to come to Trieste was extended by Christopher Zeeman, a prominent professor of mathematics from the University of Warwick in the United Kingdom. Zeeman had asked me to attend and ultimately to co-direct an ICTP school and workshop on dynamical systems. Little did I realize that I would remain involved in these activities, which would take place every two or three years, for decades. I stepped down as an organizer of this event in 2003. Others have assumed responsibility for its organization and the school and workshop have remained on the ICTP activities' calendar to this day.

I was immediately enthralled with Salam's intelligence, charm and commitment. Who wouldn't be? Here was a scientist from a poor, politically troubled country who had become an internationally renowned physicist. His accomplishments were a reflection of his undeniable genius. In all my exchanges with Salam, I was always struck as much by Salam's humanity (and humility) as I was by his professional achievements. He had a magnetic yet genial personality that proved every bit as alluring as his gilded résumé.

Salam always conveyed a deep interest in my research – not only for the intellectual challenges it presented but also for how I managed to successfully pursue my interests while living and working in a developing country. More generally, he displayed – both in words and deeds – an intense focus on the plight of scientists in developing countries and the steps that needed to be taken to improve their working and living conditions.

Salam never wavered in the ultimate goal that he sought to achieve: He wanted to create nothing less than an inviting research environment in every country that would allow scientists to remain at home once they had completed their education and training. At some point, he hoped that all scientists would find their own countries to be hospitable places in which to advance their careers and contribute to the welfare and material well-being of their societies.

From this perspective, I shared much with Salam. While I may not have known of him until I met him, his idealism and sense of purpose touched my core beliefs.

Pillars of thought

Intellectually, my ideas about the role of science in society had been shaped by the great French mathematician and philosopher, Henri Poincaré, whose writings inspired me. Working at the turn of the late 19th and early 20th centuries, Poincaré ranks among the world's greatest scientists. Although not nearly as well known to the public as Albert Einstein, in the fields of mathematics and the philosophy of science he continues to hold a position of esteem similar to Einstein's position in physics.

Poincaré was also a gifted writer and one of the great popularizers of science in his time. His book, *The Value of Science*, not only examined the relationship between physics and mathematics in ways that stimulated the thinking of scientists for generations to come, but also presented a powerful argument for the positive impact of science on society. This impact, according to Poincaré, not only resonated in terms of improved material well-being but also in helping to nurture a state of rational thinking that ultimately benefitted and enriched all people.

I found Poincaré's focus on the value of science, instilled in his writings, to be closely akin to Salam's ideals. Yet, Salam's expressions of these ideals ultimately proved to be even more captivating to me. Unlike Poincaré, Salam was not a historic figure but a living example of an individual seeking to advance virtuous goals in today's world. His work, moreover, proved instrumental in helping to move discussions from the realm of intellectual abstraction to real progress on the ground both for scientists and societies in the developing world.

For me, Poincaré offered a set of intellectual constructs on how science operated both as fundamental element and driving force in the world of ideas. Salam, in contrast, spoke about how science could make the world a better place for all people. ICTP – and subsequently TWAS – were designed as organizational tools to help transform this worthy goal into reality. There were street-smart aspects to Salam's efforts that I found particularly attractive.

I was born in 1940 in Uberaba, Brazil, a city of several hundred thousand people in the state of Minas Gerais. The city serves as an important trading centre for agriculture, livestock and manufacturing. It has also become an important centre for higher education. I received my primary- and secondary-school education there before moving to Rio de Janeiro when I was sixteen. I graduated from the University of Brazil (which subsequently became the University of Rio de Janeiro) in 1962, majoring in engineering.

With a growing interest in pursuing an academic career in mathematics, I was accepted and enrolled in graduate school at the University of California at Berkeley in the United States. There I earned a doctorate in mathematics in 1967 under the supervision of the eminent mathematician and Fields Medallist Stephen Smale. My major field of study was dynamical systems.

This was time of great social and political unrest in Brazil. The military exercised authoritarian rule over the country following a coup in 1964. I decided to spend an additional year at Berkeley as a postgraduate, hoping that the political situation would improve. I returned home in 1968, motivated by a desire that I later would learn echoed Salam's: I wanted to help my country to strengthen its scientific capacity and improve its economic and social well-being by nurturing and then drawing upon its indigenous scientific expertise.

Thinking big

Brazil's military rule, which would last two decades, was a period of repression, largely marked by lost opportunities. Nevertheless the government did provide channels for scientists to pursue their careers. I was offered a position as a professor of mathematics at the University of Rio de Janeiro. But I decided to pass on the opportunity in favour of a post with the Institute of Pure and Applied Mathematics (IMPA).

The decision carried personal financial risks. At the time, IMPA hired people under annual grants, not formal contracts. Relying on soft money, we weren't sure we would have a position – at least a paid position – from one year to the next. This financially precarious situation would remain in place from 1968 until 1975 when my appointment as a professor was confirmed and I was given a permanent contract.

What IMPA lacked in job security, it more than made up for in flexibility. It had fewer bureaucratic restrictions, provided greater time for unfettered research and offered brighter prospects for career advancement than other places. On the downside, of course, there was a greater possibility of being let go.

In retrospect, my decision to take a post at IMPA was a reflection of my desire to pursue my dreams even if it meant sacrificing the comfort and security afforded by working under the umbrella of more established institutions. Again, I was motivated by personal goals that I unwittingly shared with Salam, who often urged – indeed beseeched – colleagues and associates to "think big". It was among his favourite expressions. He used it not only to convey his own convictions but also to encourage others to join him in his cause.

When I met Salam, he didn't have to do much to convince me of this sentiment. It was something that had influenced my thinking from the earliest stages of my career. For me, an important corollary to this belief was to take chances and to be audacious.

As was true of ICTP, I knew nothing about TWAS until I was introduced to the Academy in the mid-1980s during one of my periodic visits to ICTP to organize a school and workshop on dynamical systems. My first reaction to the Academy was based on my past experience. Simply put, TWAS was another creation of Salam, so it had to be a worthwhile venture.

Yet my interest in TWAS was also piqued by my growing experience in fostering scientific capacity and excellence in the developing world. In the early 1980s, Salam had asked me to go to Pakistan to discuss strategies for improving the state of research and training in mathematics there. It was an eye-opening experience. I was not only exposed to the state of mathematics in another developing country, but I also learned that, while analysts spoke broadly of science in the developing world, each country faced its own distinct challenges in trying to build scientific capacity and excellence. This insight would serve me well, especially during my tenure as president of TWAS.

In the early 1980s, I was also invited to participate in a study sponsored by the United Nations Educational, Scientific and Cultural Organization (UNESCO), focusing on the reconstruction of science in the developing world. In meetings prompted by this initiative, I met many other scientists and scholars who shared my interests.

In addition, I had acquired a much deeper understanding of science policy options in the developing world thanks to my election to the Brazilian Academy of Sciences in 1972 and my increasing administrative responsibilities at IMPA, where I eventually became director, serving in that capacity from 1993 to 2003.

My first official engagement with the Academy occurred in 1988 when I was awarded the TWAS Prize in mathematics. The ceremony took place at the TWAS general conference in Venezuela. I was not only pleased and honoured to receive the award, but I also had an opportunity to witness the impact that TWAS's international conferences were having on science in the developing world. These events, another brainchild of Salam, have truly been unique gatherings. There is no other venue where the full range of scientific research is placed on public display, and there is no other venue where the state of science in a developing country – in this case Venezuela – is examined in such depth and in the presence of both scientists and public officials from throughout the developing world.

Over the past two decades, science conferences in the developing world have increased exponentially, both in number and range of topics. This trend, which we all welcome and all hope will continue, has curbed the singular importance of the TWAS general conference.

Nevertheless, the Academy's flagship event remains extraordinarily important to science in the South and it continues to set the standard by which the impact of other conferences is measured. The historical significance of TWAS general conferences as a model for others cannot be overestimated. The conferences' frameworks have been emulated, both nationally and internationally.

In the years following the TWAS conference in Venezuela, I became more involved in TWAS activities, largely by serving on various committees. This proved to be a valuable experience. It gave me an opportunity to get to know both TWAS and its history. It introduced me to the primary challenges driving the Academy's agenda. It enabled me to examine the measures that were being taken to address these challenges. And it made me aware of the wealth of talent and commitment to be found among the Academy's membership. My knowledge of TWAS has been deeply rooted in my experience as a rank-and-file member. This experience has served me well during the years that I held leadership positions at the Academy.

It was during this period that I also forged a strong friendship with Mohamed H.A. Hassan, who has served TWAS so ably for more than three decades, most notably in his former capacity as executive director.

The Academy has also benefited from the compelling and extraordinary ability of its presidents, beginning with Salam and continuing with the presidencies of José I. Vargas, C.N.R. Rao and now Bai Chunli. I consider each a colleague and friend, and I have been humbled to be placed in their company.

Yet all those who know the Academy would agree that TWAS would not be what it is today without the uncommon efforts of Mohamed. He has been the force that held the Academy together through good times and bad. He has moved TWAS forward through his unmatched administrative talents, his ability to forge fruitful partnerships with individuals and institutions that have been vital to the Academy's success, and his skill in motivating the Academy's able and dedicated staff.

Throughout the 1990s, my involvement with TWAS remained fairly constant. I was an active member of the Academy, eager to play my part in helping to ensure that it continued to advance its goals.

Global reach

At the TWAS meeting in Iran in 2000, the Academy officially elected C.N.R. Rao, who had been serving as interim president, to succeed José Vargas, as president. At the same time, it also elected a new Council. In anticipation of this change in the Academy's leadership, I was asked if I would be interested in being a candidate for secretary-general.

I was intrigued by the opportunity. It offered me yet another chance to think big and to be part of something that was larger than my own personal ambition. It also represented an opportunity to work closely with C.N.R. Rao, who, through my affiliation with TWAS, I had come to admire not only as one of the world's preeminent scientists but also as an impassioned advocate for building scientific capacity in the developing world. Over time, Rao would also become a close friend.

At the time of my election to be the Academy's secretary-general, I truly believed that TWAS's leadership, guided by Rao and Hassan, was in excellent hands and that the Academy was well positioned to continue to advance its bold and innovative agenda. Serving in this capacity, I thought, would be an exhilarating experience. It would place me in the company of individuals with proven ability to effect change and give me a front-row seat in an institution with an enviable track record for success. From my involvement with the Brazilian Academy of Sciences, I knew what an academy was designed to do. I was aware of the objectives that it sought to achieve and the important role that it could play in society. TWAS pursued the same goals as the Brazilian Academy of Sciences, but it did so on a much larger scale. The global reach of TWAS intrigued me as well.

Vargas has contributed greatly to the Academy's success. It is no easy task to follow in the footsteps of an iconic figure. His efforts not only allowed for a smooth transition from Salam's presidency. He also ably addressed some of the Academy's most compelling challenges, through initiatives designed to secure and stabilize funding and to strengthen the administrative framework of TWAS, especially by creating closer ties between the Italian government and UNESCO. Vargas helped to raise TWAS's profile in policy circles both in the South and North. That, in turn, helped to expand the Academy's impact and increase the prospects for additional funding from foundations and donors such as the Swedish International Development Cooperation Agency (Sida).

While the Academy continued to make progress on a variety of fronts, many challenges and goals remained fairly constant over its first two decades: to honour and promote excellence in research; to provide support for scientists from the developing world, and especially for young scientists; to expedite South-North and South-South cooperation in science among individuals and institutions; and to seek ways to put science to work to facilitate science-based development.

Because the challenges were constant, and because TWAS was achieving some success, the Academy was able to continue its work for many years without having to reinvent itself.

Ongoing journey

Yet the journey still is not complete – and it may never be. No country – either in the North or South – can afford to ignore scientific capacity building for long if it hopes to nurture a secure and prosperous society. On this matter, there is no North-South divide.

From certain perspectives, very little has changed since the Academy was established: A significant number of countries in the South continue to lag in scientific capabilities (by TWAS's account, 81 countries fit this description); women scientists still struggle to attain equality among their peers; young scientists still receive insufficient support even as their numbers grow; and the ties between science and society, while strengthened when compared to the past, still remain distant, especially in developing countries. And we continue to face financial challenges. At this moment, the global crisis that began in 2008 still is felt acutely in some developing countries. But we have faced such crises before, and we have never given up our commitment to TWAS's goals.

Yet from other perspectives, a great deal has changed. A number of countries in the South – led by Brazil, China and India – have made enormous strides in building their scientific capacity. In a growing number of research fields, they now rank among the world's leaders. Support for universities and research centres, including graduate and postgraduate fellowships, have increased substantially, providing hope that there will be an adequate supply of researchers in the future to continue – and expand – upon the encouraging course that has been set over the past several years. The voices of women have become both more numerous and prominent in the research community, although they continue to be largely absent from leadership positions. And, perhaps most importantly, political leaders throughout the developing world now recognize the impact that science has on society and the critical role that it plays in poverty reduction and economic growth. When you look at these trends, you cannot help but think that major aspects of Salam's dream have been realized. We have indeed 'thought big' and, as a result, we have indeed achieved big things.

All of these developments – both positive and negative – were well established at the time I was elected secretary-general. That is why I embraced the Academy's agenda, as outlined in its strategic plans, and why I worked so closely with Rao, Hassan and TWAS's Council and membership to implement it. That is also why I have envisioned my tenure as president as a logical extension of Rao's tenure as president. The seamless nature of the Academy's work – together with the shared vision of its leaders – have been among the Academy's greatest attributes.

An agenda for growth and strength

Most organizations must constantly change to stay current in a rapidly changing world. In contrast, Salam (as he had done with ICTP) created such a forward-looking institution in TWAS that it simply had to remain true to its ideals – and to execute its programmes effectively – to continue to be relevant.

In effect, over the course of the Academy's history, the world has edged closer to its agenda and vision. In the process, others have discovered how valuable the Academy can be in advancing goals that have become essential to global well-being. Not surprisingly, the Academy's profile and presence in the world of international science and development has steadily grown.

Here are the core aspects of TWAS's agenda over the past decade:

- The Academy has sought to strengthen its finances and place them on a more stable footing. Adequate and secure funding, of course, is a prerequisite for strong and enduring programmatic initiatives. The Italian government, in 2003, redefined its funding for TWAS from a temporary to a permanent contribution; the Academy's endowment fund, launched on the occasion of its 10th anniversary, now exceeds USD12 million; and foundations, aid agencies and private companies, ranging from Sida to the global technology firm, Lenovo, in China, are accounting for an increasing portion TWAS's programmatic funding.

- The Academy has redoubled its effort to promote science in all countries and has taken steps to become more inclusive of small developing countries with lagging scientific capacities. TWAS has made a special effort to elect scientists in countries that currently have only a few TWAS members (or no members at all) and it has implemented a programme to fund research groups in the least developed countries (LDCs).
- The Academy has sought to provide greater support to young scientists to help ensure that scientific capacity building continues in the South so that critical social and environmental challenges can be met. To advance this goal, the Academy has established a TWAS Young Affiliates programme for scientists younger than 40, who are selected for five-year terms. Being named an affiliate helps to raise the recipients' profiles within the scientific community both at home and abroad. It also gives recipients opportunities to participate in TWAS activities, including attending the TWAS general conference. The Academy has also significantly expanded its PhD and postdoctoral fellowships programmes, thanks to the support of the Italian government, Sida and a growing number of developing countries. The programmes, which began with 30 grants in the mid-1980s, now offer more than 500 grants on an annual basis.
- The Academy has sought to extend greater duties to its five regional offices. The offices bear major responsibility for the selection of the Young Affiliates, organize regional conferences for young scientists and help identify scientists worthy of consideration for membership to TWAS. The effort to decentralize the Academy's activities, largely by empowering TWAS's regional offices, is likely to continue and, in fact, accelerate in the future. The ability of the regional offices to assume greater responsibilities will constitute one of the prime measures of the Academy's

future success. It will also serve as a reflection of growing scientific capacity in countries across the South.

- The Academy has strengthened its efforts to support women scientists. It has elected an increasing number of women as members and has sought to include women on the TWAS Council. The Academy has also provided support for the Organization for Women in Science for the Developing World (OWSD) ever since the idea was conceived at a conference sponsored by TWAS and the Canadian International Development Agency (CIDA) in 1988. (Until 2011, the organization was called the Third World Organization for Women in Science, or TWOWS). With more than 4,000 members, OWSD is the world's largest organization for women scientists. Thanks to funding from Sida, OWSD also oversees one of the world's largest postgraduate training fellowship programmes for women scientists from sub-Saharan Africa and the LDCs.
- TWAS has played a critical role in strengthening South-South cooperation in science. In fact, it is fair to say that TWAS has been one of the chief architects of fostering collaboration among scientists in developing countries – a strategy that has gained currency in both the scientific and development communities over the past three decades, due in some measure to the efforts of the Academy. We see this collaboration in terms of TWAS's successful PhD and postdoctoral fellowship programmes, which now involve a growing number of centres of excellence in developing countries.

In January 2013, at the end of my second term, the mantle of leadership went to Bai Chunli, the president of the Chinese Academy of Sciences. As in the past, the transition went smoothly. We were all delighted to leave the Academy in such capable hands. Bai is not only a world-class scientist; he also has a wealth of managerial experience and ties to the international scientific community that should serve the Academy well in the years ahead.

Accounting for success

I am proud of the progress that TWAS made during my presidency. Building upon the success of my predecessors, the Academy's membership grew from 839 at the end of 2006 to 1,060 at the end of 2012. Even more importantly, the membership has never been as accomplished and diverse as it is today. For example, over this same period, the proportion of female membership rose from 6% to slightly over 10%.

The range and scope of the Academy's activities have also grown. The number of postgraduate and postdoctoral awards increased by more than 40% and the number of research grants by nearly 90%. TWAS's fellowship programmes are now among the largest and most respected in the world.

The international scientific organizations hosted by the Academy in Trieste – IAP, the global network of science academies, and the InterAcademy Medical Panel (IAMP) – have grown considerably. They are engaged in wide-ranging initiatives for the promotion of global science that add lustre to the organizations themselves as well as to TWAS by virtue of the Academy's relationship to them. TWAS, meanwhile, is now one of the world's most respected voices for the promotion of science and development. Institutions both in the South and North eagerly seek TWAS as a partner in their desire to improve both scientific capacity and, more generally, economic and social well-being throughout the developing world.

All those who have been actively involved in this journey realize how much the success of TWAS has been due to the Academy's continuing emphasis on collegiality and collaboration – among both its individual members and institutional partners. The environment that TWAS has nurtured – the friendships it has created, the sense of purpose it has instilled, and the bold notions that it has engendered and campaigned for – are intangible yet profound elements that have helped to account for the Academy's success.

These factors – immeasurable yet irreplaceable – are a powerful source of what I like to call the magic that has led TWAS to become one of the most important organizations for promoting scientific capacity building and science-based development in the South. I am equally convinced that these factors will guide TWAS to even greater heights in the years ahead.

It is not just the Academy's enviable track record, but also the way it goes about advancing its mandate that gives me confidence in the Academy's future. TWAS's goals and strategies have withstood the test of time. I am sure that they will continue to serve the Academy as well in the future as they have in the past.

On the horizon

In this spirit, I would like to conclude with an excerpt from a speech by Abdus Salam, which he gave at the Academy's first general meeting, held in Trieste, in the summer of 1985. The excerpt is a confirmation of the importance of science to development and, consequently, a clear sign of the importance that Salam placed on TWAS as a force for scientific and social progress in the developing world.

"We tend to forget," Salam said, "that it is the science of physics that brought about the modern communications revolution and gave real meaning to the concept of one world and its interdependence.

"We tend to forget," he continued, "that it is the science of medicine that brought about the penicillin revolution.

"We tend to forget it is the sciences of chemistry and genetics that brought about fertilizers and the Green Revolution.

"And we tend to forget that it is to these sciences that the developing world must turn to for resolutions of its present problems."

Since those early years, TWAS's interests have grown to encompass all the sciences – agriculture and engineering, earth science, space science and the social sciences. Our work and our mission have always sought to expand, to have a broader impact.

Today, 30 years after the creation of TWAS and the first general meeting, Salam would be delighted to know that the role of science in the developing world has never been so prominent, and that TWAS has remained at the forefront of efforts to advance science in South – not just for the sake of science, but for the sake of society.

That's why, on the occasion of the Academy's 50th anniversary, less than 20 years from now, I foresee the possibility that Salam's ultimate goal – even if it has not yet been reached – may be clearly visible on the horizon: a world in which each country has a robust scientific community comprised of skilled researchers capable of participating as equal partners in the global quest to place science in the service of all humankind.

Making it happen

Mohamed H.A. Hassan

Mohamed H.A. Hassan served as TWAS's executive secretary and then as its executive director for more than a quarter-century. The success TWAS achieved under his management has made his name synonymous with the organization.

Hassan was born in 1947 in Elgetina, Sudan. He received his primary and secondary schooling in Sudan, which included a two-year post-secondary school certificate from the University of Khartoum in 1965. He was subsequently awarded scholarships to pursue an undergraduate degree in mathematics, which he obtained from the University of Newcastle in the United Kingdom in 1968. He earned a master's degree in advanced mathematics and a doctorate degree in plasma physics from the University of Oxford in 1969 and 1973.

After his return to the University of Khartoum in Sudan, Hassan rose in the academic ranks from a lecturer to a full professor and dean of the school of mathematical sciences. Dissatisfied with his career prospects, especially his ability to engage in international research, in 1974 he travelled to Trieste, Italy, to visit the International Centre for Theoretical Physics (ICTP).

It was there that he met Abdus Salam. The encounter took place not by appointment but by chance – and it would alter his life. Several years after they met, Salam invited Hassan to come to Trieste to help launch the Third World Academy of Sciences. Nearly three decades later, he would still be in Trieste serving as the Academy's executive director.

Hassan has not only witnessed the development and growth of TWAS from a front-row seat, but he has also been front-and-centre in shaping the Academy's development. Working closely with the Academy's presidents – Abdus Salam, José I. Vargas, C.N.R. Rao and Jacob Palis – he was responsible for executing the strategies that have made the Academy a global voice in the international scientific community.

As the executive director of TWAS, he helped to devise and implement a number of innovative scientific capacity-building initiatives in the developing world. These initiatives include a merit-based competitive research grants programme, an awards programme honouring scientific excellence in the South, one of the developing world's largest PhD and postdoctoral fellowship programmes, and the development and organization of TWAS general conferences that have become signature events for examining the state of science in the South.

Hassan retired from the post of TWAS executive director in 2011, but he remains the Academy's treasurer. He is currently co-chair of IAP, the global network of science academies, and chair of the Council of the United Nations University (UNU). In addition, he sits on the Board of Trustees of Bibliotheca Alexandrina *in Egypt, the Council of Science and Technology in Society (STS) Forum in Japan, the Board of the International Science Programme (ISP) in Sweden, the Board of the Science Initiative Group (SIG) in the United States, and the International Advisory Board of the Centre for Development Research (ZEF) in Germany. He previously served as president of the African Academy of Sciences and chair of the Honorary Presidential Advisory Council for Science and Technology in Nigeria.*

Hassan has published widely in his major fields of research: mathematics, plasma physics, fusion energy, wind erosion, and dust and sand transport in dry lands. He has also written articles on science, technology and economic development in the South that have been published in such leading international journals as Nature *and* Science.

He is a member of the Colombian Academy of Exact, Physical and Natural Sciences, Royal Academy of Overseas Sciences in Belgium, the Academy of Sciences of Lebanon, Academy of Sciences of South Africa, and African Academy of Sciences, Cuban Academy of Sciences, Islamic World Academy of Sciences and Pakistan Academy of Sciences. His honours include Comendator, *Grand Cross, and the National Order of Scientific Merit in Brazil, and Officer, Order of Merit of the Italian Republic.*

In the following essay, Hassan describes TWAS's tribulations and triumphs from his unique perspective as a keen observer and, more importantly, as a key participant in the Academy's development from its earliest days to the present.

***

Working as a young mathematician at the University of Khartoum in Sudan in the early 1970s, I certainly knew about Abdus Salam. Virtually every mathematician and physicist did. Salam was someone whom scientists, especially young scientists in the developing world, revered. As a preeminent researcher and an enthusiastic and untiring advocate for science, he was an esteemed figure – the personification of what a scientist could and should be.

He was not, however, someone whom I ever thought I would meet. To say the least, it was difficult for me to imagine sitting across a table from him and speaking informally about our professional and personal lives.

But that is exactly what happened in 1974, when during a trip to Italy on behalf of my father's business, I decided to travel to Trieste to visit the International Centre for Theoretical Physics (ICTP), the renowned physics research and training institute that Salam had created in the early 1960s.

I remember leaving Trieste's train station in the late afternoon, after travelling more than five hours from Milan, and climbing onto a bus for the 15-minute ride to the ICTP campus in Miramare on the outskirts of the city. The view was not much different than today. The languid green-blue Adriatic Sea traced the bus route on the left and the grey-laced, limestone-studded hills – the *Carso* – angled upward on the right. It was a clear day, and straight ahead I could see the snow-capped Alps on the horizon.

I arrived at ICTP at about 5:30 in the afternoon. The staff had left for the day. The campus was virtually empty. I entered the main building, ascended the staircase to the second floor and turned towards the director's office.

The door was ajar. To my surprise, I saw Salam sitting at his desk. He looked up, said hello and motioned for me to come in. He then asked who I was. That's how the most important meeting in my life began.

Home and abroad

At the time, I was a lecturer of mathematics at the University of Khartoum in Sudan. I had received a doctorate degree from Oxford University in the UK in 1973 and had returned home to begin what I hoped would be a rewarding career as a teacher and researcher.

However, like so many other young researchers in the developing world at that time, I soon felt a discouraging sense of isolation. Without access to current journals and with no scientist working in my area of research to speak to, my research languished. The situation was sapping my enthusiasm and deflating my career prospects.

Growing doubts about my faltering career had prompted my trip to Italy. My father was a successful businessman in Sudan. He had asked me to go to Italy to purchase machinery for his soap factory in Khartoum. I was fortunate because working with my father was always an option if I did not succeed as a university research professor. My side trip to ICTP in Trieste was designed to determine what options might be available to help me overcome – or at least mitigate – the challenges I faced at the University of Khartoum. In my mind, discussions at ICTP would likely be my best, last hope to remain in science.

Salam and I spoke for more than an hour that afternoon. We then continued our conversation the next morning. I was astonished by the amount of time that Salam gave me. Surely, I thought, he had more important things to do. I subsequently discovered that his deep concern for others was not only a reflection of the person he was but also one of his greatest attributes. He never forgot the difficulties that he had faced as a young researcher: the professional isolation he felt in Pakistan, the loneliness he endured in the UK, and the agonizing choice he had to make between his dedication and love for his family and his dedication and love for science.

His ability to weave his own personal narrative – replicated in the experience of others – into a broad tapestry about trends in science in the developing world helps to explain why he succeeded on so many levels. It also helps to explain why he continues to be a prominent figure in discussions about science in the developing world nearly two decades after his death. His story was a compelling one, filled with facts, figures and, most importantly, people. Despite his enormous intellect, Salam spoke from the heart.

Drawing on his detailed knowledge of university conditions throughout the developing world, Salam told me I had two options if I wanted to continue my career in the world of science instead of joining my father in the world of business.

To pursue my scientific ambitions, he said I either could change my field from plasma physics to high-energy physics and transfer to the physics department at the University of Khartoum, which had retained an excellent faculty and reputation in high-energy physics, or I could continue as a member of the mathematics faculty. The latter, he cautioned, would likely create additional obstacles that would make it even more difficult to advance my career.

A lifeline to global science

Regardless of the path I chose, Salam strongly suggested that I should apply to visit the Centre on a regular basis as an ICTP Associate. His power at ICTP, of course, was considerable – indeed unchallenged. His words, as a consequence, were more than words. They meant that my selection was assured.

Being appointed an ICTP Associate enabled me travel to Trieste several times over the course of the next few years, breaking my isolation at home and introducing me to a larger network of scientists abroad.

Equally important, Salam's intervention at this critical juncture in my career steered me from my father's business ventures. My own stubbornness, meanwhile, kept me in the mathematics faculty. Having been trained largely in the field of mathematics, I was determined to be a mathematician. For Salam, the latter was a minor consideration. More importantly from his perspective, I would not be lost to the business world. I would remain a scientist.

Over the next few years, my encounters with Salam would be sporadic, largely confined to lunchtime encounters during my stays in Trieste and attending his presentations and lectures at the Centre. Salam would rarely miss an opportunity to introduce a workshop or a prominent guest at ICTP, and young scientists at the Centre would rarely miss an opportunity to attend these events.

In 1982, I joined several other faculty members in convincing the University of Khartoum to give Salam an honorary degree. Salam travelled to Sudan for the event. This gave me another chance to see and speak with him. While day-to-day activities at the university remained dreary and difficult, my ties to ICTP – both at home and abroad – gave me a lifeline to the international research community.

The birth of TWAS

It was during Salam's visit to Sudan that he initially told me about his plans to create an academy for scientists from the developing world. He mentioned that he had first presented the idea at an informal lunchtime gathering of eminent scientists from the South at the Pontifical Academy of Sciences in Rome, Italy, which had taken place the year before. He said that the enthusiastic response of his colleagues, which included a declaration voicing unanimous support for the concept, had prompted him to try to turn the idea into a reality.

Salam then surprised me by asking if I would be interested in helping him launch the academy. I said "yes". Salam was not someone you often said "no" to. Besides, I owed him a great deal. His intervention had saved my career.

Yet, the truth is that I didn't have an inkling about what I was getting myself into. I barely knew what a science academy was, and I knew even less about what such institutions did or why they might be needed.

A few weeks after Salam had returned to Trieste, he sent me a letter asking if I would come to ICTP for six months to help organize what he referred to as the "foundation meeting" of the Third World Academy of Science (TWAS).

Having previously said yes to the idea, there was no way I could say no to the invitation to help launch the organization. The truth is that I didn't want to say no. So, it was off to Trieste. Little did I realize that this six-month stay would turn into a commitment spanning nearly 30 years, largely defining my career in science.

The makings of an academy

The first order of business was to learn about the mission, purpose and structure of other science academies – as much as I could, as fast as I could – so that we could apply the experience of science academies, some of which were hundreds of years old, to TWAS. During those early months I spent a great deal of time reading about academies. In effect, I developed my own tutorial about these institutions, closely examining three academies in particular: the Pontifical Academy of Sciences, the UK's Royal Society and the Academy of Sciences of the Soviet Union.

All were comprised of eminent scientists. All were designed primarily to serve their members. All promoted science as an integral part of society. And all honoured scientific excellence and accomplishment as the centrepiece of their efforts.

Although they all shared similar goals, they differed significantly in how they conducted their business.

The Pontifical Academy of Sciences was, in many respects, the most similar in structure and purpose to what we hoped to create at TWAS. It was international in scope and examined broad issues of importance not only to its membership but also to science and society. Moreover, it was linked to an entity, the Vatican, with responsibilities and sensibilities that extended well beyond the borders of a nation state. It was truly an international science academy with a broad and compelling mission. That was exactly what TWAS aspired to be.

The Royal Society, founded in 1660, was one of the world's oldest science academies. It was deeply respected around the world. In fact, it was viewed as a model by many other academies, especially in the West. The UK's most eminent scientists were members and the Society's reports helped to shape both the national research agenda and global discussions on science.

The Academy of Sciences of the USSR, which served as a model for socialist countries, operated under a different administrative framework than science academies in Western Europe and much of the Americas. In addition to honouring and supporting eminent scientists, it was directly responsible for managing the USSR's largest and most prestigious research institutes. In the mid-1980s, its institutes employed more than 250,000 scientists.

TWAS was structured as a hybrid of each of these academies. Yet it would also establish a niche as a one-of-a-kind institution.

Like the Pontifical Academy of Sciences, TWAS would assemble a broad international membership and engage in issues well beyond the conventional roster of in-country concerns that largely characterized national academies. Like the Royal Society, TWAS would be interested in issues related to global progress and would harbour ambitions to gain a presence on the world stage. And, while TWAS had neither the resources nor the intentions to be directly involved in the management of scientific facilities, like the Academy of Sciences of the USSR, it would be concerned about the administration of day-to-day scientific activities, seeking to draw lasting links between scientific activities and economic policies.

Forty-two and counting

The most immediate challenge that TWAS faced was expanding its members beyond the 42 eminent scientists who had founded the Academy in 1983. They belonged to the world's most prestigious academies of sciences such as the *Accademia Nazionale dei Lincei* in Italy, the French Academy of Sciences, the Royal Society in the UK, the Royal Swedish Academy and the US National Academy of Sciences.

A key element of our thinking during the first years of TWAS was that the future of the Academy – both in the near- and long-term – depended on quickly increasing its membership while ensuring that those chosen as members represented the developing world's best scientists.

However, we soon discovered that identifying scientists in the developing world who were worthy of election to TWAS would be no easy task. While developing countries constituted more than 80% of the world's population, relatively few world-class scientists worked in the South, and those who did were dispersed over a large geographic area spanning virtually every continent. The task proved even more difficult because scientists, including those doing excellent work, were often not well known within their own countries, let alone by scientists elsewhere. Communications in this pre-Internet age were difficult and slow. Interactions among scientists in the developing world were often scarce and sporadic.

Given these obstacles, we devised a number of interrelated strategies in our efforts to build the Academy's membership. For example, we asked the founding members to identify potential candidates. We reached out to visitors at ICTP who may have known scientists back home who had yet to gain the recognition they deserved. We contacted science academies not only in the North but also in the South, especially the largest and most active science academies in the developing world – in Argentina, Brazil, China and India, for example – for suggestions about potential candidates. And we identified eminent scientists in the North with ties to the South.

Our relentless efforts to expand the Academy's membership ultimately paid off. Within five years of its launch, the Academy had 200 members.

TWAS was being rapidly built – one worthy scientist at a time – through a meticulous mission that relied on just a few signposts for its success. To illustrate how difficult it was to fill the academy's ranks, it should be noted that, while the majority of the founding members were born in the developing world, half of them lived and worked in the North.

Gaining respect

The second major challenge facing TWAS during its critical formative years was the difficulty it encountered in winning recognition from international organizations.

Many advocates for international science, including those active in the International Council for Science (ICSU), did not see the need for an organization like TWAS. Their reasoning was that respected institutions were already in place to speak for international science. Confusion and duplication of effort, they contended, might follow if additional organizations, with largely the same responsibilities, were created.

Critics also maintained that science was a global enterprise and dividing scientific interests into developed and developing world spheres could jeopardize the universality of the enterprise. And, finally, there was a sense that an institution dedicated to the interests of scientists in the developing world would have neither the personnel nor resources to abide by the scientific community's commitment to excellence.

The arguments were not much different than those raised during debates surrounding the creation of ICTP in the early 1960s. At that time, critics contended that the training of physicists from the developing world could take place at research institutions that existed in the developed world – at Cambridge in the UK and Princeton in the USA, for example. Why go through the time and expense of creating duplicative institutions, they reasoned. Critics also questioned whether research centres dedicated to scientists from the developing world would be capable of conducting world-class science.

Gauging success

TWAS's challenge was to illustrate how it could add value to global efforts to build scientific capacity and how it could serve a useful role in applying that knowledge to critical development issues, especially in the developing world. Moreover, it needed to address these challenges without compromising the scientific community's abiding dedication to excellence.

Today, TWAS works closely with ICSU and other international scientific organizations and is widely recognized as a key player in efforts to promote and advance both scientific capacity and scientific excellence in the developing world. However, during those early years, it was not clear – at least to wary observers – whether TWAS could shoulder any meaningful responsibilities in the arenas where it intended to operate.

In retrospect, part of the problem was that international scientific institutions did not have a sufficient number of scientists from the developing world among their membership to ensure that the voices of the South were being heard. Viewing the situation from the vantage point of history, it is clear that scientific issues of critical importance to the South were not being adequately addressed. The problem was not due to wilful neglect. Instead it was the consequence of a lack of awareness.

Another key aspect of the problem was that international scientific institutions worried that additional organizations would divert scarce financial resources to places that did not have the wherewithal to fulfil their avowed commitments.

On this front, the issues were twofold. First, we had to illustrate how TWAS would complement and not replicate existing efforts at scientific capacity building. Second, we needed to prove that the Academy was up to the task.

These challenges served as the basis of the third critical test that the Academy faced during its formative years: how to devise and implement a series of concrete programmes to deal with issues of critical importance to scientists in the developing world that had not yet been addressed – or that been addressed inadequately – by other scientific institutions. In this effort, the Academy did not have to thread a needle within a dense fabric of competing and overlapping programmes. Instead it had to envision a different pattern of support for a constituency that others had failed to acknowledge.

To advance this goal, TWAS organized international conferences that were meant, among other things, to promote South-South cooperation at a time when meetings focusing exclusively on the concerns of scientists from the developing world were rare events. The Academy also created an awards programme for scientific excellence in the South that was designed to honour scientists who had attained virtually no recognition for their accomplishments. And it sponsored research grants for scientists, particularly young scientists, from the South, who were hard-pressed to find funds.

Trifling support for science on the part of national governments and the absence of private foundations in the South left scientists in most developing countries desperate for money to sponsor their research. The TWAS research grants programme, supported with funds from the Italian government and later the Swedish International Development Cooperation Agency (Sida), has been the primary plank in the Academy's efforts to build scientific capacity. No other programme more clearly represents what Salam hoped to accomplish through TWAS, and no other programme more clearly conveys the success it has achieved.

None of the Academy's early initiatives, which have come to define TWAS over time, could be found at other international scientific organizations. In this sense, TWAS was complementing the work of others by serving a constituency that had been largely ignored. In the process, the Academy was helping to build scientific capacity worldwide, a contribution to international science that the global scientific community quickly came to recognize and appreciate.

In this regard, the Academy was also careful to complement, not duplicate, the work of ICTP, whose long-standing initiatives for scientific capacity provided the template for many TWAS programmes. For example, while ICTP focused its capacity building efforts on physics and mathematics, the Academy sought to encompass the full spectrum of scientific disciplines, including biology, chemistry, environmental science and medical research. Later in its evolution, TWAS opened its membership to social scientists.

Similarly, ICTP's Training and Research in Italian Laboratories (TRIL) programme brought scientists from the developing world to research institutions in Italy – an effort that proved to be a sterling example of North-South cooperation in science. TWAS, in turn, grafted the TRIL concept onto a programme for South-South cooperation in science by partnering with preeminent research institutions in the developing world to provide research and training opportunities for scientists from countries with lagging scientific capabilities. TWAS's South-South programme followed the same path as the TRIL programme. But it did so on a global scale while tilting the axis from North-South to South-South collaboration.

From dream to reality

Identifying centres of scientific excellence in the South during TWAS's early years posed enormous challenges. Convincing scientists from developing countries to travel to these institutions for training and collaborative research proved even more difficult. In the 1980s, most scientists throughout the developing world desired to travel to the North for collaboration (a preference that persists to this day, but to a far lesser degree). The developed world was their clearest pathway to success. They neither trusted the level of education and training they would receive in developing world institutions nor believed that pursuing South-South collaboration in science would enhance their career prospects.

As a consequence, the Academy's initial efforts at South-South cooperation in the 1980s resulted in less than 30 participants annually. Today, the Academy, together with its partners, sponsors more than 500 South-South research and training opportunities each year. Moreover, the number of scientifically rising developing countries offering training and research opportunities to scientists in less scientifically proficient developing countries has expanded from Brazil, China and India to include Malaysia, Mexico, South Africa and a growing number of other developing countries that have strengthened their research institutions to become increasingly proficient in science.

In the 1980s, South-South cooperation was largely an abstraction – or, when viewed from Salam's perspective, a dream. Thirty years later, it is a reality – one of the defining elements of the new paradigm that is emerging in global science. TWAS's successful efforts to promote South-South cooperation represent one of the Academy's most significant contributions.

The period between 1983 and 1992, the first decade of the Academy's existence, was a time devoted to fundamental principles and precedent-setting activities. The focus was on how to gain recognition and respect within the international science community, a goal that was largely dependent on the quality (and to some extent the size) of our membership. TWAS, after all, needed a critical mass of TWAS Fellows if it hoped to have a presence in international science circles and a collective voice that would be loud enough to have its message heard.

In addition, TWAS concentrated on how it could assist developing countries in their efforts to build their scientific capacity and boost their economies. Here the challenge was to develop a set of effective programmes that met a need that wasn't being addressed elsewhere.

During this first decade, the Academy also tried to establish enduring frameworks for South-South and South-North collaboration that would help shed light on the scientific research that was taking place in developing countries (at least among some developing countries) and that would assist in overcoming the stigma associated with science in the South. In this sense, TWAS was not only a capacity-building programme; it was also a confidence-building programme.

And finally, the first decade of TWAS's existence was also a period in which the Academy laid markers to illustrate that broad-based reform would require the involvement of many institutions. Indeed it would require an extensive network of scientific and economic development organizations working together. It would also necessitate reaching out to the policy community to ensure that national and international policies acknowledged and promoted science as an integral part of the development process. Such concerns explain why Salam and TWAS were so eager to play a pivotal role in the creation of the Third World Network of Scientific Organizations (TWNSO) and the Third World Organization for Women in Science (TWOWS).

Years of challenge

TWAS's next decade – the years following the Academy's 10th anniversary celebration held in Trieste in 1993 and culminating with the Academy's return to China in 2003 for its 20th anniversary – were the most difficult in TWAS's history. The Academy faced three formidable challenges during this period.

First, it had to secure a framework for sustainable core funding that would guarantee reliable support for its administration and programmatic activities. Second, it had to confront the loss of its founding president and guiding light, Abdus Salam. And, third, it had to navigate a transfer in administration and management responsibilities from the International Atomic Energy Agency (IAEA) to the United Nations Educational, Scientific and Cultural Organization (UNESCO).

Looking back, I can now say that each of these challenges also presented opportunities that the Academy responded to in ways that ultimately made the organization even stronger. At the time, however, I was not nearly as sanguine about the Academy's future. In fact, I sometimes thought TWAS faced a daunting set of problems that placed its future well-being – indeed its very existence – at risk.

The Academy responded to its funding challenges in two ways. With the help of its friends and allies, it redoubled its efforts to change the Italian government's funding scheme from a voluntary financial contribution, subject to annual reviews, to a permanent financial contribution written into Italian law. At the same time, in 1996 it launched an endowment fund that would depend solely on contributions from developing countries. The fund's initial goal was set at USD10 million. When that target was reached in 2008, on the occasion of TWAS's 25th anniversary, it was reset to USD25 million. Today, contributions to the endowment fund exceed USD12 million.

The Italian government's generous support for TWAS represents an excellent example of North-South cooperation. The initial funding from the Italian government brought TWAS into existence and allowed the Academy to launch a broad spectrum of interrelated activities that helped to build its reputation in the international science community. The Italian government's subsequent decision to make its financial contribution permanent, finalized in 2003, ensured that TWAS would be able to continue its good work by creating a firm financial foundation upon which it could strengthen and expand its activities.

The financial crisis that the Academy faced in the 1990s was matched by a crisis in leadership. In the late 1980s, following an unsuccessful effort to become UNESCO's director-general, Salam fell ill to a chronic and debilitating Parkinson-like disease. Although his intellect remained razor-sharp to the end of his life, the disease stifled his energy and, over time, even his ability to communicate. In 1994 Salam announced that he would be stepping down as president of TWAS. Finding a replacement would prove both delicate and difficult. Salam was irreplaceable. Moreover, as a practical matter, who would want to follow in the footsteps of such a revered figure?

Fortunately, after much cajoling (he had to be convinced he should do it), José I. Vargas, Brazil's minister of science and technology and an active figure in international organizations, especially UNESCO, agreed to become the next president of TWAS. Vargas assumed the leadership of the Academy in 1996 and would remain at the helm until 2000.

Vargas brought many positive attributes to the position. As a prominent figure in scientific and development circles in Latin America, he helped raise the profile of the Academy not just in Brazil but also throughout the Americas. By the early 1990s, TWAS was well-known among the scientific communities in Asia, Africa and also in Islamic countries. But the Academy had a much lower profile in Latin America and the Caribbean. Vargas was instrumental in helping TWAS to become more active and visible throughout the region.

Similarly, he helped the Academy to develop collaborative activities with UNESCO, where he was chair of the executive board and subsequently Brazil's representative to the organization. Under his guidance, UNESCO not only served as TWAS's administrator, overseeing personnel and budgetary matters, but also partnered with the Academy on activities designed to build scientific capacity in the South.

Equally important, Vargas helped to strengthen TWAS's engagement with issues that extended beyond the sphere of scientific capacity building for individual scientists to questions of broad science policies, ranging from overall investments in research and development, to the relationship of science to economic development, and to the critical role that science plays in education and innovation. It is not that these issues hadn't been discussed before. What Vargas did, however, was to draw them to the centre of the Academy's agenda.

Emerging strengths

Throughout its history, TWAS has exerted a level of energy and impact that has seemed to exceed its budget and staffing. The Academy's ability to punch above its weight could be linked to a number of factors.

The Italian government has provided a consistent source of funding for TWAS. Just as important, its confidence in the Academy has enabled TWAS to pursue a nimble and aggressive agenda. Affiliation with the UN and UNESCO has not just facilitated its partnerships with international organizations, but has also bolstered recognition for the Academy across the globe. Meanwhile, ICTP, which operates under the auspices of both UNESCO and IAEA, has provided office space and administrative support – without which TWAS may not have survived its early years.

A 1991 agreement between Italy, UNESCO and TWAS was a major advance that truly allowed the Academy to come into its maturity.

Under the agreement, TWAS would work under UNESCO's financial oversight and personnel rules. And it would continue to rely on ICTP for matters like procurement, payroll and office space. But the agreement gave TWAS new responsibility for managing its own personnel and budget. Within this framework, TWAS could forge new partnerships with other scientific organizations, reach out to new funders, and build an endowment fund.

The new paradigm, with a stable operating framework provided by trusted partners, gave TWAS the freedom to grow into an influential force in global science.

2003: a watershed year

If the inaugural decade of the Academy can be characterized as the years of firsts and the second decade the years of challenge, then the third decade can be viewed as the years of stability and expansion. This period, the most successful in TWAS's history, began with the success of the TWAS 20th anniversary celebration in China and largely continues to this day.

In many ways, 2003 was a watershed year for the Academy. Not only did the conference in China receive international media coverage (confirming the increasing importance of the Academy and its partners), but it also reflected the Academy's growing ties with the developing world's scientific communities. The conference in China has been followed by other equally successful conferences, for example, in Argentina, Egypt, Brazil, Mexico, South Africa and India. As in China, the countries' presidents have opened most conferences. Each conference, moreover, has detailed the state of science in South and each has showcased the increasing strength of South-South cooperation.

The conferences have represented only the most visible aspect of the Academy's role in building scientific capacity and fostering South-South collaboration in science. Progress has also been reflected in the expansion of the Academy's endowment fund. It has been demonstrated in the commitment that China, India and Brazil have made to the Academy's fellowship programme. It has been displayed by TWAS's ability to launch a successful grants programme for research groups in the developing world's least science-proficient countries, in addition to continuing its funding for individual scientists. It has been shown, as well, by the Academy's increasing efforts to lend support to young scientists through its Young Affiliates programme, its grants and awards programmes for scientists under the age of 40, and its sponsorship of regional conferences for young scientists largely designed to foster exchange among recipients of the Academy's grants.

Such efforts illustrate that TWAS no longer needs to concentrate exclusively on immediate challenges, but can think – and act – broadly about how to nurture the next generation of scientists in the South. Indeed, helping young scientists is now a major aspect of the Academy's agenda.

All of these advances were guided by two central figures who have been largely responsible for the Academy's growing strength and visibility in the global scientific community: C.N.R. Rao, who served as president from 2000 to 2006, and Jacob Palis, who served from 2007 through 2012.

During his tenure, Rao helped guide the Academy through an extraordinary period of growth by reformulating its agenda to focus additional attention on advancing scientific excellence – whether in the selection of its members or in the awarding of grants to young scientists. He also sought to reverse the growing South-South divide which has been unfolding between developing countries that are rapidly building their scientific capacity and those that are lagging behind. And he led the drive to devote additional attention to demographic groups that have yet to fully participate in the developing world's scientific capacity efforts. These groups include young scientists and female scientists in addition to scientists living and working in the developing world's poorest countries.

Rao also continued to expand the Academy's role as a critical voice for examining the state of science policies in the developing world and for exploring whether all segments of society – government, the private sector, nongovernmental organizations and scientific institutions and agencies – were playing their part in helping to build scientific capacity and applying that capacity to address vital social and economic needs. Finally, his tenure also witnessed the creation of five regional offices in Brazil, China, Egypt, India and Kenya. It is fair to say that under Rao's direction, TWAS became the voice of science in the South.

Palis' legacy lies in expanding the Academy's agenda in ways that have enabled it to become a truly global player in the international science community. It is only fitting that the Academy at the conclusion of Palis' tenure elected to change its name from the Academy of Sciences for the Developing World, which it had adopted in 2004, to The World Academy of Sciences.

During Palis' tenure, the Academy's fellowship programmes continued to grow. TWAS currently offers more than 500 fellowships each year. Hosting countries include Brazil, China, India, Malaysia, Mexico, and South Africa, making it one of the largest fellowship programmes in the developing world.

Under Palis' leadership, TWAS's regional offices also began to assume greater responsibilities for the Academy's activities, marking the first significant moves toward decentralization. These offices, with financial support from TWAS, now help to identify promising young scientists for potential selection as TWAS Affiliates, organize regional conferences for young scientists and engage in discussions about policy issues of importance to the region.

All of this growth and change took place as the Academy continued to focus on scientific excellence, as exemplified by the eminent scientists who were elected to TWAS, the high level of the candidates applying for Academy fellowships, prizes and grants, and the launch of such initiatives as the Ernesto Illy Trieste Science Prize in 2005. The prize quickly became a highly prestigious honour for the developing world's most distinguished scientists. Lenovo, China's multinational computer technology company, assumed sponsorship for the USD100,000 prize in 2013, as President Bai Chunli was coming into office. The TWAS-Lenovo Science Prize has not only enabled the Academy to continue to forge strong links with the corporate world, but the collaboration is now taking place with one of the South's largest and most prestigious firms.

Progress and challenges

Over the past three decades, the Academy has served both as an insightful observer and an agent of change during an historic time for science in the South. Indeed the growth of TWAS has parallelled the growth of science in the developing world.

Think of the state of science in the developing world in 1983. Think of the state of science now. Think of the state of TWAS in 1983 – the limited range of its activities, its budget, its staffing, its visibility. Think of the organization's broad range of activities and reputation now.

Salam's visionary goal of attaining scientific excellence in all countries, which seemed like a pipe dream 30 years ago, is much closer to reality today. Yet compelling challenges continue to stand in the way of progress.

Because of its success, TWAS is well-positioned to play a critical role in meeting these challenges. Yet, it must continue to evolve to remain a prominent and influential voice in a world that is changing at an unprecedented pace.

First, I believe that the Academy should take steps that would enable its regional offices to shoulder additional responsibilities, especially in the administration of TWAS's fellowship programmes. The rapid expansion of this activity has made it a centrepiece of the Academy's work, requiring the secretariat's staff in Trieste to devote a significant portion of its time to the initiative. The prospects for future programme growth – the number of fellowships could conceivably increase from 500 today to 1,000 over the next decade – will only add to the secretariat's workload.

The only way that the Academy can take full advantage of this unique opportunity for growth is to give the regional offices primary administrative responsibility for the programmes – for overseeing the appointment of students and arranging for travel and lodging.

This would not only allow the programme to run more efficiently, but it would also free the secretariat in Trieste to do other important work, including developing and implementing a postdoctoral programme to provide ongoing support and guidance to fellowship recipients at critical transition points in their careers.

Second, TWAS must continue to build its membership base. The Academy's strength and reputation depend largely on the quality and size of its membership. This is as true today as it was during TWAS's early years.

I envision the Academy increasing from 1,000 members today to 2,000 members by its 50th anniversary in 2033. That would mean the Academy will have expanded at an average rate of 40 scientists a year – in effect, adding scientists at an annual pace that would be roughly equal in size to its founding class. The growing reputation of the Academy, combined with the increasing number of excellent scientists now working in the South, places this goal well within reach.

While the Academy's membership continues to rise, it must also make a concerted effort to reduce the average age of its Fellows, which is now 70. And it must also seek to add a substantial number of women scientists to its ranks.

Over the next 20 years, I would like to see the average age of scientists in TWAS decline by at least 10 years to 60 (or even less), and for women scientists to represent at least 30% of our membership compared to 10% today.

Again, given current trends in science in the South, these goals should not be dismissed as wishful thinking. Rather, they should be embraced as eminently achievable.

The number of scientists in the South is rapidly increasing due to rising investments in science by governments and the growth of such initiatives as the TWAS research and fellowship programmes. South-South cooperation, moreover, is on the rise, helping to improve scientific collaboration and capacity. Meanwhile, the voice of women scientists is becoming stronger and more influential due in part to the work of groups such as the Organization for Women in Science for the Developing World (OWSD). TWAS is proud to have played an instrumental role in OWSD's development and success. All of this bodes well for science in the South – and for the ability of TWAS to extend the size and diversity of its membership.

Third, TWAS must continue to expand its efforts to be a leading voice on policy issues related to science and development in the South. The Academy has sought to engage the policy community on a broad range of development issues in which science plays a critical role – through statements and reports analysing key policy concerns, including strategies for building scientific and technological capacity, ensuring safe drinking water and expanding the production of renewable energy. It has also sought to organize forums that bring together the scientific, development, finance and diplomatic communities to discuss issues of common interest.

The time has come to systematize these efforts. If the secretariat were to be relieved of its duties in administering the fellowship programmes, it could then devote more time to expanding the Academy's role in the science policy arena. This would help shift TWAS's focus from a management agency to a global think tank without compromising its efforts to assist scientists and scientific institutions.

TWAS could then take steps to strengthen its relationships with other international organizations, particularly within the UN system, which have special mandates to address critical global issues but do not have the same deep relationships or credibility with scientific communities in the South that TWAS has nurtured over the past three decades. These agencies include UNESCO, the United Nations Development Programme (UNDP), the Food and Agricultural Organization (FAO) and the World Health Organization (WHO).

Working in partnership with these organizations, TWAS could for example help in advancing global agendas to improve science education, devise effective strategies for sustainable development, enhance food production and minimize the threat posed by the spread of infectious diseases.

At the same time, TWAS, by concentrating an increasing portion of its efforts on policy issues, could work more closely with scientific communities and institutions in the North. Nearly 15% of TWAS members are employed by scientific institutions in the developed countries and many TWAS members – whether working in the North or South – have collaborated with colleagues around the world. One of the Academy's greatest strengths, in fact, is its international network that radiates out from the developing world to spread across the entire globe.

Bridging divides

The history of TWAS includes numerous examples of scientists and scientific institutions inviting the Academy to join in efforts that have been led by the North. It may now be time for TWAS to take the lead and invite colleagues and institutions in the North to join the Academy in efforts to address some of the most critical issues we face as a global community.

TWAS could thus act as an institutional bridge uniting scientists and scientific institutions in both the North and South to better serve our increasingly interdependent world. TWAS's interactions with merit-based science academies – which have been strengthened over the past decade through the Academy's close ties with IAP, the global network of science academies, whose secretariat is located in Trieste – have reinforced TWAS's credentials as a key player in promoting science-based sustainable development at the global level.

Broadening the Academy's agenda to achieve a greater global presence and impact will, of course, require additional revenues. TWAS's current annual budget, which totals approximately USD5 million (including more than USD2 million from the Italian government and almost USD2 million from Sida) is both a reflection of the generosity of others and a measure of how far TWAS has come from its earliest days marked by severe budget constraints and uncertainties.

The Academy is thankful to be on a firm financial footing. Yet, if it wants to do more, it will have to secure more funding.

TWAS should seek to increase its annual budget to USD25 million, which would include USD10 million for the activities of the central office and USD3 million for the activities of each of the five regional offices. A budget of that magnitude, if sustained over the next 20 years, would go a long way towards enabling TWAS to complete the journey that Salam set out for the Academy some three decades.

Even in light of the difficulties of the global economy currently faces, I believe securing such a level of funding is possible. TWAS no longer has to prove itself. The Academy is a well-respected institution in international science. Its place at the table is no longer in doubt. Scientists in the developing world eagerly seek to gain membership into the organization (as do scientists from the developed world). International organizations often turn to TWAS for its perspectives on critical issues related to both science and development, and welcome the Academy's participation in collaborative projects. Leading political figures speak in praise of the organization.

The challenges remain formidable. But TWAS is in a much stronger position to meet them than it has ever been. For this we have much to celebrate as we mark the Academy's first three decades – and for this we have much to look forward in the years ahead.

When I first became involved in TWAS more than three decades ago, I had no reason to believe that the Academy would accomplish so much. Yet, when I look towards the Academy's future, I believe that the goals Salam envisioned for TWAS are in reach. For that reason, I believe it is the future, even more than the past, which makes this anniversary period so important.

When looking back on the occasion of the Academy's 50th anniversary, observers may well be speaking of how TWAS, in the years following its 30th anniversary, took the steps that were necessary to enter a new phase of relevance and importance to truly emerge as The World Academy of Science for the advancement of science in developing countries.

Minding the gaps

Ana María Cetto Kramis

Ana María Cetto Kramis has played a leading role in global efforts to improve the status of women scientists in the developing world. She has also been a tireless advocate for science and development in the South.

Cetto Kramis served as founding vice-president of the Third World Organization of Women in Science from 1988 to 1998 (TWOWS has since changed its name to the Organization for Women in Science for the Developing World, OWSD); vice-president of the Commission on Physics for Development at the International Union of Pure and Applied Physics (IUPAP) from 1999 to 2001; vice-president of the Committee on Science for Developing Countries (COSTED) at the International Council for Science (ICSU) from 1990 to 2001; and secretary-general of ICSU from 2002 to 2008. Between 2003 and 2010, she was the deputy director-general at the International Atomic Energy Agency (IAEA) in Vienna, Austria, where she headed the Technical Cooperation Department. She worked for the IAEA when the organization won the Nobel Peace Prize in 2005 and was a member of the Executive Committee of the Pugwash Conferences when it won the Nobel Peace Prize in 1995.

Cetto Kramis has written numerous books and articles on topics ranging from quantum mechanics to the role of science and technology in development. She is the recipient of many honours, including the International League of Humanists' Golden Award, the Mexican Physics Society's Prizes for the Development of Physics and for Research in Physics, and the Sor Juana Inés de la Cruz Award from the National Autonomous University of Mexico (UNAM). She was named Woman of the Year in Mexico in 2003.

Cetto Kramis was born in Mexico City in 1946. She earned bachelor's, master's and doctorate degrees in science from UNAM and a master's degree in biophysics from Harvard University.

She is currently a research professor and lecturer at UNAM's Institute of Physics, working on the foundations of quantum physics. In addition, she serves as founding president of Latindex, an online information system for Latin American and Caribbean scientific journals; director of UNAM's new Museum of Light; and a member of the steering committee of the International Year of Light, an initiative led by the European Physical Society designed to celebrate and enhance public understanding of the central role of light in the modern world. The year-long celebration is being held in 2015.

In the following essay, Cetto Kramis examines how her involvement with TWAS, which has largely taken place through positions she held with other institutions, has helped to both raise the status of women in science and enhance the role of scientists in the developing world in global organizations. Whether the focus has been on gender or geography, a quest for equity in international science has been the hallmark of her career.

***

I first learned about TWAS sometime in 1986 from senior colleagues at the National Autonomous University of Mexico's (UNAM) Institute of Physics, where I was young researcher in physics. Several prominent Mexican physicists at the Institute, including Marcos Moshinsky, had been elected to the Academy, and news about their membership was publicized both on campus and in the media.

Before then, I was not aware of TWAS's existence. I did, of course, know about Abdus Salam, the Academy's founding president, and the force behind the creation and growth of the International Centre for Theoretical Physics (ICTP). Salam was a famed figure among scientists in the developing world, especially after he had received the Nobel Prize in 1979.

ICTP, which had been launched in the mid-1960s thanks largely to Salam's tireless efforts, was well known among scientists in the South. This was especially true among physicists and mathematicians who had journeyed to the Centre in Trieste, Italy, to attend workshops and conferences. By the 1980s, thousands of scientists had done so.

An old men's club?

As for TWAS, it was launched in the early 1980s and had yet to gain a presence among scientists in the developing world. In fact, when I first heard about TWAS I found myself largely uninterested. For me, it seemed like all other science academies, only on an international scale.

Academies in the 1980s were largely old men's clubs – grey-haired scientific fraternities that left little room for women. I thought TWAS fit that mould, and I didn't think it had the capacity to be anything different in the future.

My views of TWAS would change after I was invited to participate in the conference, "The Role of Women in the Development of Science and Technology in the Third World", held in October 1988 in Trieste. Discussions about organizing a conference on the status of women scientists in the developing world – or, more precisely, on their lack of status – had taken place the previous year at TWAS's second general meeting in Beijing. The need for a women's conference was evident to anyone who cared to look. Virtually no women scientists were present at the assembly in Beijing. That shouldn't have come as a surprise since there were virtually no women among the members of TWAS.

The women's conference in Trieste proved to be a landmark event. It was the first time women scientists from both the North and South assembled to talk about issues of common concern and to discuss among themselves how they might devise strategies to overcome their second-class status. The conference influenced the careers of a number of women scientists, including my own. I think it is fair to say my involvement in international activities would have taken a different path or, at least, moved at a slower pace if I hadn't been there.

All told, about 250 women scientists participated in the conference. There were also representatives from prominent international organizations such as the United Nations Educational, Cultural and Scientific Organization (UNESCO) and the International Council for Science (ICSU). Salam assembled the participant list by asking a wide range of contacts, including ICTP alumni and TWAS members, for the names of both prominent and promising women scientists who they thought should be invited to attend. I was a researcher at UNAM and was fortunate enough to be on that list.

A movement is born

The eventual outcome of the meeting was the creation of the Third World Organization of Women in Science (TWOWS), a trail-blazing organization that was designed to assist women scientists in much the same way that TWAS had been designed to assist all scientists in the South. Yet, given the gender imbalance in science, for TWAS, that meant focusing largely on the challenges faced by male scientists. In contrast, TWOWS was founded to close the gender gap by lending a helping hand to women scientists – or, to state it differently, to fill a gap that was present in TWAS.

Like the Academy, TWOWS sought to place the vital concerns of its membership on the agenda of international science. More generally, it gave female members of the scientific community, who had largely been relegated to an insignificant role, an important stage on which to express their opinions and have their voices heard and counted.

The creation of TWOWS was no easy task. In fact, the debate at the founding conference in Trieste focused, in part, on whether women scientists should seek to improve their status within existing organizations such as ICSU and TWAS or to establish a separate organization dedicated solely to issues related to women.

There were compelling points to be made on both sides of the argument. It was possible, for example, that TWOWS could find itself just as isolated among international science organizations as individual women scientists had found themselves in university departments, research institutes and science academies within their own countries. Would it be better for women to work within the existing structures of science, however limiting they may be, or to strike out on their own in an effort to spark fundamental reforms? Were the voices and concerns of women more likely to be heard inside or outside the conventional channels of science?

In a sense, the discussions leading to the creation of TWOWS evolved around the question of where women scientists would be least marginalized and therefore most empowered in their efforts to reform the *status quo*. This was a discussion that many women scientists thought was well worth having.

Some may be surprised to learn that there were considerable disagreements among women about the forces driving discrimination and isolation in the scientific workplace. This was especially true in Latin America, the region where I live and work.

In discussions with my female colleagues from the region, I discovered that, while not a majority, a sizeable number of female scientists – sometimes the most successful ones – believed that their good fortune was due solely to their individual talent and initiative and, conversely, that failure was largely a personal responsibility. They contended that they did not need nor want an organization for women scientists designed to provide special dispensations and allowances for the intensely competitive world in which all scientists work. They also believed they could address the challenges of balancing the demands of their careers and families by hiring outside help to care for their children or by turning to their relatives for support. Moreover, they contended that their families provided a degree of solace and a sense of identity and worth that no professional organization, however well-meaning, could replicate.

Cultural mindsets and practices, especially in Latin America, seemed to reinforce the conventional attitude that in science you make it on your own (with the help, of course, of your family and mentors). This highly individualized pathway to success, in the eyes of TWOWS's female critics, applied to both men and women.

In contrast, women on the other side of argument, who were led by TWOWS's founding president Lydia Makhubu, maintained that the abysmal state of women scientists in the developing world illustrated that it was time for a fresh start and new ideas – and that the best way to advance the goals of women scientists for fair and equitable treatment was through the creation of a new, independent organization.

It was their sentiments that prevailed. But the opinions of others were heard and considered during the creation and formative years of the organization.

TWOWS may have helped to launch a movement to improve the status of women in science by marshalling their members to make the case for the need for change. Yet the movement was by no means monolithic, and a vigorous debate accompanied the decision-making process. Perhaps the most significant factor was that the debate, through the forum that TWOWS provided, took place among the women scientists themselves. That, in itself, was illuminating and empowering.

At the beginning, TWOWS stood alone as an institution dedicated solely to women's issues in science. While the organization focused on women scientists in the developing world, the involvement of women scientists from the North was encouraged.

This should not be surprising. Unlike the North-South gap in science, which was marked by huge differences in scientific capabilities among regions, the gender gap in science was a global phenomenon affecting women scientists in developed and developing countries alike.

Working conditions, of course, varied considerably between the North and South for both male and female scientists, and few women scientists in the North would have traded places with their female colleagues in the South. Yet prevailing attitudes toward gender, to a large degree, were a global phenomenon. Critical issues related to woman scientists simply did not conform to the same paradigm as the North-South divide in science overall. That is why TWOWS was, in significant ways, a truly global institution from the start.

Following the women's conference in Trieste, I became actively involved in the creation of TWOWS. Salam asked me to join the committee responsible for drafting the organization's constitution, which I gladly did. After the constitution was approved, I was elected vice-president of TWOWS for South America and the Caribbean. Playing a significant role in the organization's formative years proved to be both an exhilarating and rewarding experience. In addition to crafting the constitution, we needed to devise statutes, solicit members, seek funding, develop an agenda and build the organization's reputation.

TWAS proved to be an instrumental ally in our efforts. TWOWS had no budget during these years and its survival depended on the Academy's beneficence. The truth is that TWOWS's fragile existence could not have been sustained without TWAS's commitment and generosity. For this,

TWOWS owes a great deal of gratitude to both Abdus Salam and the Academy's executive director, Mohamed H.A. Hassan, who kept the organization afloat when the concerns of women scientists were not high on the agenda of international scientific or development communities.

No organization likes to depend on others for its survival. That, in and of itself, is a sign of weakness and vulnerability. But if that is the reality an organization faces, it would indeed be fortunate to have friends like TWAS. This is a part of the Academy's history that has often been eclipsed by its other accomplishments. Yet TWOWS is an important part of TWAS's legacy.

Today, many scientific organizations in both the North and South devote considerable attention and resources to issues of importance to women in science. There is now a consensus that both national and global progress on many fronts – not just in science but also in economic development, public health and environmental well-being – depends, to an ever-increasing degree, on the growing number of women who are active not only in scientific research but also in higher education, management and policy.

A persistent gender gap

A good deal of progress has been made in advancing the status of women in science – at least compared to the situation at the time of TWOWS's creation. Universities, research institutes and international organizations currently offer a large and growing number of scholarships and fellowships for young women scientists. Postgraduate positions and mentorship programmes for women researchers early in their careers have been created in some institutions. Female role models are more numerous and prominent.

In some fields – for example, medicine and environmental sciences – women scientists may approach or exceed 50% of the profession within their nations. The increasing number of females enrolled in undergraduate and graduate school, including in many scientific disciplines, suggest that these percentages are likely to rise in the future. It should also be noted that a number of developing countries, including several in Latin America, have achieved greater gender equality in science than countries in the North.

Yet, this does not mean that the issues that gave rise to TWOWS have been resolved. Far from it. While women scientists in both the North and South have made great strides in achieving equality in the classroom and laboratory, the gender gap has not been entirely eliminated. This is especially true in such disciplines as mathematics and in my own field of physics, where males continue to dominate. And while gender trends in teaching and research have, for the most part, been encouraging, this is less the case when it comes to positions of management and leadership, where too often the old men's club continues to prevail.

The ongoing gender gap in science has consequences since the demands traditionally posed by family and motherhood present special challenges for the careers of women scientists (and indeed all women in the workplace).

Appointing more women scientists in decision-making positions would make it more likely that such issues would be addressed with greater forthrightness and urgency – although as my experience during the early years of TWOWS suggests, we should not expect women to speak with one voice on gender or, for that matter, on any other issues.

As women gain more authority, however, that will allow policies to be increasingly shaped by those who are directly affected by the decisions. The process itself would be viewed as more equitable, and the decisions, in turn, would be accepted as more legitimate – not only by women scientists but also, I believe, by the majority of male scientists who are fair-minded and realize that times have changed.

TWAS's instrumental role

The same concerns for access and equity have been at the centre of much of the work that I have done with international organizations such as ICSU, UNESCO and IAEA. But in these forums, the challenges have focused on how to provide access and equity for scientists – both male and female – from the developing world. Here again, TWAS has proven to be an invaluable ally.

One of the things that struck me when I first became involved with international organizations in the early 1980s was the absence of scientists from the developing world within their ranks. For me, this seriously compromised their claims of having a global reach. In fact, in my mind, while these organizations may have been international in the sense that representation came from multiple countries, they weren't global at all. Indeed, representation from 80% of the world could hardly be found. Northern administrators and researchers led these organizations, which, not surprisingly, pursued agendas that were almost exclusively attuned to Northern interests and concerns. Multinational, yes; global, no.

Equally important, while organizations came to recognize this shortcoming, it was clear that they had neither a sufficient appreciation for the work being done by scientists in the developing world nor sufficient ties to scientists in the South to make significant progress on this front. Their hearts may have been in the right place and their rhetoric may have struck the right chords. Yet the organizations were not adequately grounded in scientific communities in developing countries to nurture true partnerships with the growing number of excellent researchers working there.

That is where TWAS came into play.

I experienced first-hand how valuable TWAS's involvement could be while serving as a member of the advisory committee of ASCEND21 (Agenda for Science and Development into the 21st Century), a conference that was held in 1991, just one year before the Rio Earth Summit. TWAS helped to ensure that the concerns of the developing world were taken into account, not just by the presence of Northern scientists who were pursuing research agendas devoted to issues of importance to the developing world, but also by the presence of researchers working there. The ensuing reports (16 in all) were co-authored by scientists from the developing world. These publications helped to frame discussions not only at ASCEND21, but also at subsequent meetings dedicated to sustainable development, including the Rio Earth Summit.

I again witnessed the value of TWAS's involvement in the activities of international organizations while working on the organizing committee of the World Conference on Science in 1999, which was co-sponsored by UNESCO and ICSU. There was great interest in having significant participation from scientists from the South, and TWAS was asked to assist in this effort. The Academy helped to ensure that the goal of broad participation was met by drawing upon its extensive network of eminent scientists, a network based largely on its membership roster.

The involvement of TWAS, in fact, was instrumental in making the conference a success. It not only increased the presence of developing-world scientists at the event but also helped to shape the conference's agenda, assuring that issues of vital concern to developing countries – for example, scientific capacity building, brain drain and South-South cooperation – were given an adequate airing. I think it is fair to say that the conference marked an important milestone in global science by bringing scientists from the developing world to the discussion table in ways that they hadn't been brought before.

Since then, other international science organizations, councils and academies have been created, and in these places the presence of scientists from the developing world has increased significantly. Part of the reason lies in the growing scientific capabilities of a rising number of developing countries, and part of the reason lies in the proactive efforts of international scientific organizations to engage scientists throughout the developing world.

Both TWAS and ICSU have established regional offices and both have pursued efforts to link science to economic development and critical social and environmental challenges. UNESCO has pursued a capacity-building initiative that seeks to improve the science policy infrastructure in developing countries by working closely with science and other ministries. These efforts have both recognized and advanced scientific capabilities in the South in ways that have made capacity building more sustainable over time.

In a similar vein, during my tenure at IAEA from 2003 to 2010, the Technical Cooperation Department (TCD), which I directed, transformed its programmes into true collaborations by encouraging member states to devise initiatives that fit into their broader strategies for scientific and technological capacity building. The member states themselves provided funds to help advance largely self-directed efforts to build strong foundations for science-based development and to address critical energy, food, environmental and health challenges.

IAEA's mandate is to promote peaceful applications of nuclear science and technology to improve the social and economic well-being of the citizens in their member states. TCD's reforms were designed to make the programme less top-down and agency-directed and more demand-driven and member-state-directed so that its efforts would both reflect and address national needs.

TCD has made significant progress on this front. And because member states are now engaged in designing and implementing activities, there is every reason to believe that the advances which have been made – for example, in improving water quality, raising agricultural productivity and enhancing the diagnosis and treatment of disease – will be sustained.

IAEA recognized that traditional methods of technology transfer had failed. Simply handing down technologies from the North to the South proved an unreliable path to reform that rarely spurred broad and lasting progress. IAEA also realized that the reforms introduced would have fallen flat – indeed would likely not have been introduced at all – if not for the growing scientific capabilities of institutions within the member states themselves. Under the new partnership strategy, scientific capacity became not a goal but a process. The more capacity a nation has, the more valuable it becomes and the more likely it is that it can be expanded. The same principle has been at the heart of TWAS's efforts ever since the Academy's inception.

Identity and pride

So, where does the important progress that has taken place in scientific capacity building in the South leave TWAS as it faces a future far different than the past?

First of all, we should all recognize that the 30th anniversaries of the Academy's founding and its first General Meeting are reasons for celebration. The unprecedented growth in scientific capabilities in developing countries is one of the most significant global developments of the past three decades, and TWAS deserves a great deal of credit for being an important player in this effort.

The world has entered a new era of international science in which the distinct lines drawn between the North and South have been blurred in part and, in some cases, largely eliminated. This historic development has been reflected in the recent decisions by both TWAS and TWOWS to change their names. TWAS, in fact, has changed its name twice over the past decade: first, from the "Third World Academy of Sciences" to the "Academy of Sciences for the Developing World" in 2004, and then to "The World

Academy of Sciences for the advancement of science in developing countries" in 2012. TWOWS, in 2010, became the Organization for Women in Science for the Developing World (OWSD). The name changes are a reflection of the rapidly changing world in which we live.

I fully appreciate TWAS's pride and desire to create a name for itself that more accurately characterizes the world-class status of both its members and the organization itself – and foregoes the negative connotations historically associated with the terms "Third World" and "developing world". But I would caution TWAS not to stray too far from its roots even as the world of science moves ahead at lightning speed, bringing the North and South ever closer together.

The Academy is *of* the developing world, not just *for* it. As it gains an increasing presence on the world stage of science, the organization must never forget that it remains an institution dedicated to scientific capacity building in the South. That has been its historic role and that has been the source of its identity and, I might add, pride. As Salam said many years ago during earlier discussions about the best name for the Academy: "We are what we are, no matter what we call ourselves."

Lessons worth learning

Historically, the achievements of TWAS, OWSD and many other scientific organizations devoted to science in the South have been largely defined by Northern concepts of what it means to successfully pursue and advance science.

There was much that the South could – and indeed did – learn from the practice of science in the North – lessons about transparency, openness and excellence – that have proven to be universal in their value and effectiveness. But I think we may now have reached a point in global science where the South – and particularly women in the South – can impart important lessons in the practice of science to their counterparts in the North.

Take, for example, the way in which applications of science to development in the South over the past three decades have fuelled unprecedented rates of economic development in some developing countries. Are there lessons to be learned by the North from this experience – for instance, about long-term strategies for investment in science, about the role of government in promoting and advancing science, and about the importance of science education in primary and secondary schools?

Likewise, the call by women scientists in both the North and South for more collaboration and less competition could help create a more hospitable environment for young researchers and encourage the creation of a more diverse scientific workforce.

Such developments would be good for science and good for society. In addition, they could be led by segments of the scientific community that have historically operated on the periphery of international scientific organizations and academies but are now increasingly at centre stage. Such developments, moreover, would signal that these once-marginalized segments of the scientific community are not just 'catching up' but establishing innovative agendas worthy of study and perhaps emulation by others in both South and North.

What I am suggesting is that the reform agenda of TWAS – and other international science organizations – over the next 30 years may prove to be even more dramatic and significant than it has been over the past 30 years. It may be driven by an agenda in which the South doesn't just succeed by diligently tracking the pathways of success created by the North, but actually blazes its own pathways of success that over time gain adherents from the North.

Likewise, it may be an agenda in which women succeed not solely on the terms established by their male counterparts but rather by embracing new methods of work and collaboration. These methods could prove to be equally productive, yet more in line with the career expectations and lifestyles that scientists and non-scientists alike have projected for themselves, their families, their societies and their nations in the years ahead.

The good news for TWAS is that the Academy is as well-positioned to play as prominent a role in these potential reforms as it was in helping to lead the way in past reforms. That could make the coming decades of the 21st century as eventful for TWAS as the past three decades have been. At that point, the grey-haired fraternity that I believed TWAS to be when I first learned of the Academy nearly 30 years ago would truly have become an artefact of the past.

Opening doors

Adnan Badran

Adnan Badran was born in 1935 in Jarash, Jordan, the capital city of one of the nation's 12 governorates (administrative units), located about 50 kilometres north of Amman.

A renowned plant biologist, Badran has enjoyed a long and distinguished career as a scholar, national political figure and international diplomat. He currently serves as president of the University of Petra in Amman, Jordan, president of the Arab Academy of Sciences, and president of the Arab Forum for Environment and Development (AFED).

Over the past three decades, Badran has held a number of ministerial posts in his home country, including prime minister and minister of defence (2005), and minister of agriculture and then minister of education (1989). Since 2006, he has been a senator and chair of the Senate Committee on Education, Science, Culture and Media.

As secretary-general of Jordan's Higher Council for Science and Technology (1986-1987), Badran served as one of the chief architects of Jordan's national science and technology policies. As founding president of Yarmouk University (1976-1986), he helped lay the groundwork for the growth of higher education in Jordan. In pursuit of the latter goal, he has also been president of Philadelphia University (1998-2005).

At the international level, Badran was UNESCO's assistant director-general for science (1990–1994) and deputy director-general (1994–1998). During this period, UNESCO broadened its efforts to incorporate science as a critical aspect of the organization's activities, paying special attention to promoting international cooperation in science.

Badran was elected a member of TWAS in 1991. He served as secretary-

general from 1992 to 1998 and as vice-president from 1999 to 2003. In the early 1990s, he played a key role in the administrative transfer of TWAS from the International Atomic Energy Agency (IAEA) to UNESCO. Later in the decade, he helped TWAS forge closer ties to international organizations, most notably UNESCO and the International Council for Science (ICSU).

Trained as plant biologist at Oklahoma State University and Michigan State University in the United States, Badran has published a large number of academic articles in peer-reviewed scientific journals. In addition, he has co-authored several biology textbooks for both secondary school and university students and has written extensively on issues related to science policy, science education and international cooperation in science. He also holds several international patents, most notably on techniques for improving the storage and transport of apples, bananas and other fruit.

In the following essay, Badran describes the impact that Abdus Salam had on the development of science in Jordan. He then speaks about his involvement with TWAS during a critical period of the Academy's development in the 1990s – a time when the fledgling institution weathered a number of crises to emerge as a strong and enduring presence in the global science community. He concludes by exploring how TWAS can assist scientific communities in the Islamic region during this period of dramatic and convulsive change.

***

In 1980, Abdus Salam came to Jordan to receive an honorary degree from Yarmouk University. At the time, the university was just four years old. Nevertheless it had already attained an enviable reputation for its international outlook, having recruited faculty from Europe, Turkey and the United States and having attracted a diverse student body. The university began with 640 students. Today, it has nearly 30,000.

In the late 1970s and early 1980s, higher education in Jordan was on the upswing, and Salam wanted to lend his support to our efforts. Well before the creation of TWAS, he had been an indefatigable advocate for science in the South, travelling endlessly to highlight the work of the International

Centre for Theoretical Physics (ICTP) and to congratulate and encourage others for their promotion of science.

During his visit, Salam spoke to the university's first graduating class, outlining his compelling vision of a new world order in which all countries would enjoy levels of scientific capacity sufficient for improving the economic and social well-being of their people. He urged graduates to utilize their expertise to make this happen, and he reassured them that rewarding and meaningful careers would await them if they did.

Salam returned to Jordan in 1984 to lend additional credibility to Yarmouk University. The themes he discussed were similar to his previous visit, and he continued to laud university administrators, faculty and students for the progress they had made.

An enduring friendship

Salam's visits to Jordan in the 1980s, which are recalled to this day, proved significant for several reasons.

First, his presence served as a strong endorsement of the University of Yarmouk and, more generally, higher education in Jordan. Salam had received the Nobel Prize in 1979, just one year before his initial visit. The honour had elevated his status from a world-class scientist to a global ambassador for science, especially in the developing world. By virtue of his eloquence, passion and status, he was able to convey the critical importance of science to development in ways that virtually no one else could.

Second, it was on this occasion that Salam first met Prince Hassan of Jordan, the brother of King Hussein. The prince, who had graduated from Oxford University, shared with Salam a deep respect for the value of higher education for both individuals and society. Equally important, Salam and Prince Hassan were both dedicated to addressing the complex array of challenges in the developing world through applications of science and technology. Such efforts could only be achieved if developing countries nurtured their own expertise.

Prince Hassan and Salam quickly struck a steadfast friendship that would continue until Salam's death in 1996. The two worked closely on a number of initiatives designed to boost science-based sustainable development, especially in Jordan and, more generally, the Arab region.

Prince Hassan has remained a strong advocate for development and cross-cultural exchange, continuing his advocacy work over the past two decades. In 2013, UN Secretary-General Ban Ki-moon appointed him chairperson of the UN advisory board on water and sanitation.

And third, Salam, who was always an endless font of ideas and enthusiasm, made a compelling suggestion while in Jordan: He urged his guests to create the Petra School of Physics at the university and to convene an international conference each year dedicated to a single theme that was high on the global research agenda.

Echoing what I am sure he believed was also true for the ICTP in Trieste, he urged us to take advantage of our geography and history. I distinctly remember him saying that we should use our natural and cultural assets to our advantage. He was confident that prominent physicists would welcome the opportunity to come to Petra, if not for the conference, he noted with a wry smile, than surely for the opportunity to visit the world-renowned stone-cut architecture of this ancient trading city.

Fortunately, we followed Salam's advice, and the school that was created has achieved enormous success. The annual conferences have attracted numerous Nobel laureates in biology, chemistry, medical science and physics. They came to learn and exchange ideas. But I am sure they also came to see the stunning archaeological sites.

"A win-win-win situation"

In 1991, I was elected a TWAS Fellow, some seven years after Salam's second visit to Jordan on behalf of Yarmouk University and just one year after I had been appointed assistant director-general for science at UNESCO. My research had focused on plant biology, but I had become increasingly involved in the administration and management of science – initially in Jordan, but over time in international scientific organizations, most notably the International Council for Science (ICSU) and UNESCO.

As UNESCO's assistant director-general for science and then deputy director-general, throughout much of the 1990s I served as the organization's representative to ICTP. This gave me an opportunity to visit Trieste several times a year, including to attend ICTP's annual steering committee meeting. I had also worked closely with Julia Marton-Lefèvre, the executive director of ICSU, on a number of collaborative projects.

These undertakings would prove critical for the contributions that I was to make to the Academy during the years I was most active with TWAS. In fact, a good deal of my time and effort was devoted to trying to forge closer links between TWAS, ICSU and UNESCO.

For example, TWAS had a fellowship programme for scientists from the South. ICSU, in turn, had a fellowship programme for scientists from the North. It seemed to me that it made sense to join these two programmes. Part of the appeal was that this could be easily accomplished without jeopardizing the focus – or, for that matter, the independence – of either. At UNESCO, I subsequently created a budget for similar fellowships and aligned the programme with the existing collaborative arrangement between TWAS and ICSU.

What we did amounted to more of an alliance than a merger, allowing for a larger range of options for applicants interested in seeking research and training opportunities in institutions other than their own. The result was the UNESCO-ICSU-TWAS fellowship programme, a collaborative venture that not only added valuable resources to the effort but also raised the programme's profile.

At UNESCO, we were particularly interested in helping scientists from Africa. TWAS's long-standing involvement and unimpeachable credibility with Africa's scientific community helped UNESCO broaden its access to scientists, particularly scientists who were living and working on the continent. TWAS, in turn, garnered greater visibility and resources, and ICSU gained closer contact with colleagues in the South. So it proved to be a win-win-win situation for all the institutions.

High ideals, practical structure

In the early 1990s, ICTP and TWAS faced one of their greatest crises. The International Atomic Energy Agency (IAEA) no longer wanted to maintain lead responsibility for the administration of ICTP (and by association

TWAS). A number of IAEA board members, reflecting the perspectives of their countries' political officials, contended that neither the goals nor the activities of ICTP and TWAS were an essential part of the Agency's mandate, which concentrated mainly on overseeing compliance with the Nuclear Non-Proliferation Treaty and promoting peaceful applications of nuclear science and technology.

In the minds of the critics, the academic nature of the Trieste-based institutions resided largely beyond IAEA's focus and concerns. As a result, a sizeable number of states that were members of the IAEA, particularly in the North, argued that the IAEA should not be so prominently linked to ICTP (and indirectly to TWAS). They claimed that the Agency's contributions to both organizations, however modest, would be better spent on activities that were more in line with the Agency's primary responsibilities.

Salam was distraught about the prospects of having IAEA significantly curb – and perhaps even forego – its support for ICTP and TWAS. He fervently wanted ICTP and TWAS to remain a part of the Agency.

Many factors drove Salam's thinking. Being associated with a UN organization gave both institutions credibility and backing. The UN label also made the Italian government much more enthusiastic about continuing its financial support. In addition, Salam zealously believed in the UN mandate, especially the lofty principles that had been enshrined in the organization's charter: to promote human rights and dignity among all nations and to seek social progress and better living conditions throughout the world. Salam, not surprisingly, was convinced that ICTP and TWAS could each play a fundamental role in advancing these worthy goals.

As a more practical consideration, Salam feared that if the UN's protective umbrella were to collapse, ICTP and TWAS would no longer be shielded from the volatile political and financial environments in which they operated. The organizations' idealistic principles may have accounted for their appeal. Yet these ideals, if they were to be realized, needed a practical framework in which to operate.

The UN – along with the Italian government – provided that framework. In fact, the role of the UN and the Italian government in helping to ensure ICTP's and TWAS's well-being, in Salam's mind, were intertwined.

The response from UNESCO to Salam's predicament was both immediate and reassuring. After I explained the situation to UNESCO's director-general, Federico Mayor, he simply told Salam not to worry. He said to Salam that his house, UNESCO, was also Salam's house, and that regardless of what IAEA decided to do, ICTP and TWAS would always be welcome at UNESCO. This decision helped to open the door for a smooth transition of primary administrative responsibility from IAEA to UNESCO.

Mayor's response must have been particularly poignant for Salam. Just a few years before, Salam had actively sought the position of director-general of UNESCO, a pursuit that proved unattainable because of the unwillingness of the Pakistan government to support him in his quest.

Mayor had been elected to the post that Salam had hoped would be the capstone of his career. Now Mayor, as the head of UNESCO, was reassuring Salam that neither ICTP nor TWAS would experience any adverse repercussions from IAEA's decision to no longer be their lead administrative organization.

TWAS at 10

In 1993, TWAS held its 10th anniversary meeting in Trieste. The event has been well chronicled in articles and books written about TWAS – and rightfully so. Over the course of a single decade, TWAS had established itself as a significant force in international science widely respected for its efforts to build scientific capacity in the developing world.

Two days before the 10th anniversary meeting, Salam convened a preparatory meeting. While the anniversary meeting was largely a ceremonial event open to the public, the preparatory meeting consisted of a series of closed sessions where the organization's future was discussed in detail. The collegial, yet private, nature of the meeting encouraged participants to be less celebratory and more forthright in their views and opinions.

Think of these two days as an institutional retreat designed to provide Salam with concrete ideas on how the Academy could continue to succeed in an environment that presented a full slate of opportunities and risks. Although I had been elected to TWAS just two years before, I was invited to this conclave. At the time, I was UNESCO's assistant director-general for science.

Salam himself was at a critical juncture in his efforts to sustain the success of both ICTP and TWAS. Not only had he recently faced the difficult challenge of transferring primary administrative responsibility for ICTP and TWAS from IAEA to UNESCO, but he also had to navigate the organizations through a budget crisis in the Italian government that had threatened to halt funding for both organizations. The money eventually came through, but not before a bridge loan was secured from the Iranian government that allowed both organizations to continue to pay their staffs and bills while they awaited the release of the Italian funds.

At the same time, Salam was also confronting a personal illness that had grown significantly worse over the preceding years. This Parkinson-like disease had sapped a great deal of his energy, requiring him to call upon others for assistance. His mind, however, remained as sharp as ever, and he continued to offer a wealth of ideas designed to move his agenda forward. The chronic nature of his illness, however, meant that things would only become more difficult for him as time went on.

The public ceremonies marking the 10th anniversary of TWAS were congratulatory and joyous. These days represented the bright side of the TWAS story, and there was indeed a good deal to convey about the Academy's success and positive impact.

In contrast, the closed meetings that took place prior to the public meetings, in some sense, reflected another, less optimistic, and darker view of the Academy's current state of affairs and future prospects. I don't mean to imply the mood was pessimistic, but it certainly was more subdued.

I remember Salam noting that, perhaps now more than ever, scientists from the developing world continued to need a forum such as TWAS that would enable them to speak to one another about challenges of common concern. These challenges pertained not only to their own fields of research but also to the social and political obstacles that they all faced – including an acute lack of funding, poorly equipped laboratory facilities, excessive teaching responsibilities and an abiding sense of isolation.

Much had changed since TWAS was created, and the progress that had been made needed to be acknowledged and applauded. Yet Salam lamented that too many scientists from too many developing countries continued to work in substandard conditions with only scant prospects for productive and rewarding careers.

An unfulfilled dream

Salam observed that the barriers impeding the exchange of information between scientists in the South and the North – at least for the world's most eminent scientists – were not nearly as imposing as the barriers that existed among scientists in the South. In many developing countries, moreover, there were simply too few scientists for meaningful discussions about one's research to take place. Poor communication systems and limited travel funds further hindered such efforts. Major research questions, moreover, continued to originate in the North. That meant the research agendas in the South were largely – if not exclusively – shaped and driven by concerns in the North.

In Salam's mind, all of this meant that too many scientists in the South continued to face the same stark choice that he had faced in Pakistan nearly four decades earlier – a choice between their profession and their families, their careers and their country.

TWAS, of course, was not Salam's only contribution to scientific capacity building in the developing world. He had also built ICTP, where scientists (mostly physicists and mathematicians) from the developing world could meet to exchange ideas. The associateship programme, which he had originated at ICTP, enabled scientists to come to the Centre in Trieste on a number of occasions over several years (specifically, three times over six years). The programme was designed to encourage scientists to engage periodically with colleagues without having to permanently leave their home countries.

Despite its enormous impact, Salam acknowledged that ICTP served, in essence, as an ideal model for North-South, not South-South, collaboration in science. After all, funding from the Italian government was responsible for its success, and Trieste, Italy, was where scientists from the developing world came for research and training. Salam was more than grateful for Italy's generosity in making ICTP a success. Yet, he also wanted to create a framework that would directly engage institutions in the South as well.

Salam hoped that TWAS's scientific fellowship and exchange programmes would enable scientists to travel to universities in developing countries, making this a true South-South collaborative initiative that would be an important driver for scientific capacity building in the developing world.

Without such an effort, he was convinced that the South would face a dismal future largely devoid of original and innovative scientific research. As had been the case for centuries, virtually all research efforts – and findings – would continue to be derived (or even more likely transferred) from interactions with Northern colleagues who would largely determine the scope of the global scientific agenda.

On this score, the Academy had only begun to make headway. After 10 years of efforts, South-South cooperation in science continued to be largely an unfulfilled dream.

Salam, moreover, believed that, without home-grown scientific expertise, it would be difficult for developing countries to successfully apply technology, especially cutting-edge technology, to critical social and economic problems. As he often noted, the fundamental reason for the enormous gap in material well-being between the North and South had nothing to do with cultural and social differences. Rather, it was a consequence of differences in the mastery of science and technology.

Listen and lead

Salam was superb at deflecting opposition and continuing on his way. It is not as if he turned his back on those who resisted or opposed his ideas. He was, in fact, a sincere and careful listener, eager to engage others in discussions and skilful in the art of persuasion. His persistence in seeking to build scientific capacity in the South was as deep and unshakeable as was his intelligence and imagination in solving some of the great unanswered questions of physics, most notably uncovering the theoretical principles that explained the unification of the weak and electromagnetic forces in the universe, which had led to his Nobel Prize. All of these qualities – both intellectual and personal – help to account for his success.

Salam, of course, was acutely aware of the financial trials and tribulations that the Academy had faced throughout its brief history. He also had first-hand knowledge of how fragile funding could be for institutions such as science academies that depended on the munificence of others, most notably governments, international organizations, foundations and individual patrons, for the money they had to have to operate. He needed to look no further than his recent deliberations with both the IAEA and the Italian government, which had raised fundamental questions about the future well-being of both ICTP and TWAS.

From a much longer perspective, history had shown that academies – even those that had achieved great success – often disappeared from the scene when funding became scarce. This was certainly true in the Islamic region, which a millennium ago was home to the world's foremost centres of scientific learning and discovery. These institutions flourished from 700 AD to 1400 AD, a time in which the Islamic world dominated the world of ideas.

But internal tensions within Islam and declining economic fortunes led to a loss of support for esteemed learning and research centres in Alexandria, Cairo, Cordoba, Damascus, Baghdad, Toledo and elsewhere. This rich intellectual legacy, which had placed Islam at the global hub of science for more than seven centuries, became largely relegated to history.

Salam, of course, was well-versed in the history of these institutions and fully understood the importance of permanent funding for ensuring the health and strength of academies.

In fact, TWAS's success had been due, in no small measure, to the Italian government's generosity. Funding from Italy had sustained the Academy and allowed it to implement a series of innovative programmes that had helped to strengthen scientific capacity throughout the developing world.

Yet, while the core source of funding for TWAS had been derived from Italy, the Academy itself had been developed by and for scientists from the South. Italy, in fact, had given the Academy another gift nearly as precious as money when it also granted TWAS the independence to define and meet the challenges that the Academy deemed to be the most critical to achieving its goals.

The independence with which the Academy operated was something that UNESCO also valued in its dealings with TWAS. Even more importantly, TWAS's independence was something that UNESCO was determined to continue when it assumed primary administrative responsibility for the Academy. UNESCO believed independence was one of the primary factors accounting for TWAS's success – second only to the quality and dedication of the Academy's leadership and members.

Northern ties, Southern anchors

While Salam was dedicated to promoting science in the developing world and focused his efforts on strengthening South-South collaboration in science, he never doubted the importance of maintaining close ties with scientists from the North. That is why, he always contended, North-South and South-South cooperation in science were mutually reinforcing initiatives that helped to advance global science.

As Salam often noted, science is universal. As a result, he maintained that it would be a serious mistake for scientists from the developing world to isolate themselves from their colleagues in the North. Besides denying one of the basic tenets of science – its universality – confining Academy membership to scientists solely from the developing world would likely have the unintended consequence of marginalizing scientists from the South even further, thus undermining one of TWAS's primary goals in the process.

From the start, Salam observed, ICTP had been a gathering place for scientists from the North and South – even as it pursued its primary goal of improving the research capabilities of scientists in the developing world.

Based on this experience, no one needed to urge Salam to keep TWAS's doors open to all eminent scientists who pursued research agendas of interest to the South or who had devoted at least a portion of their careers to research issues of importance to the developing world – even if they did not live in the South. In fact, half of the Academy's initial membership roster worked in the North. TWAS ultimately institutionalized its ties to developed countries by creating an associate membership category.

This was – and remains – an important decision that ensured the Academy would be an open, not a closed, organization. TWAS has welcomed all eminent scientists worthy of recognition within the overall context of TWAS's goals.

I think this helped to set the stage for the recent name change for the institution: The World Academy of Sciences for the advancement of science in developing countries. More importantly, it has helped position TWAS to play an even greater role in global science in the future.

Yet attaining the Academy's ultimate goal – adequate levels of scientific capacity in all countries – will also require TWAS to take advantage of the positive changes that have occurred among a growing number of countries in the South.

Not alone

TWAS has never been able to accomplish its goals on its own. One of the Academy's greatest strengths has been its ability to leverage its limited resources by partnering with other entities, ranging from the Italian government to the Swedish International Development Cooperation Agency (Sida) and from UNESCO to the Packard Foundation.

Over the past decade, TWAS has turned increasing attention to collaborating with developing countries that have achieved a significant level of success in building their scientific capacity. This list, which is headed by Brazil, China and India, is not only encouraging but also growing. The primary channel by which TWAS has forged partnerships is through its fellowship programme for PhD and postdoctoral students, an initiative that parallels the strategy that marked the relationship between TWAS, UNESCO and ICSU in the 1990s.

Now is the time for the Academy to strengthen and expand its ties with scientifically proficient countries in the developing world. The goal should not only be to enhance South-South cooperation, but also to help build scientific capacity in countries that continue to lag behind. It is time to draw on the Southern anchors to help raise the level of science in those developing countries that have yet to fully participate in the advances made over the past three decades.

The Arab region, where I have lived and worked throughout virtually my entire career, would benefit significantly from such efforts. As I have chronicled in an essay in the 2010 *UNESCO Science Report* (co-authored with Moneef R. Zou'bi, executive director of the Islamic World Academy of Sciences), a growing number of Arab states had begun to increase their investments in science and technology prior to the Arab Spring in 2011. These efforts largely focused on strengthening education in existing universities or building entirely new campuses such as the King Abdullah University of Science and Technology (KAUST) in Saudi Arabia; Qatar's Education City; and the Masdar Institute in Abu Dhabi.

By some measures, these efforts have been impressive, as a number of world-class facilities have been built. Even more significantly, the number of students in the Arab region enrolled in higher education has risen dramatically, from 5.4 million in 2000 to 7.3 million in 2008.

Yet, by other measures, science has remained a low priority in many countries in the region. Average expenditures for science and technology as a proportion of gross domestic product (GDP) vary from 0.1% to 1.0%, compared to a global average of 2.5%. Only 7 of the 21 countries in the region have merit-based scientific academies. The private sector remains woefully absent from the region's research and development efforts. And jobs remain scarce, including for recent university graduates. A lack of job opportunities has often been cited as one of the reasons for the rise of the Arab Spring. Moreover, even when regional cooperation in science has taken place, it has usually been at the level of individual scientists engaged in small projects.

The Arab region, which shares many common challenges, would profit enormously from large-scale regional partnerships taking place at the institutional and national levels. Science pursued in this manner could provide benefits well beyond scientific capacity building. It could serve as a valuable instrument for bringing the region closer together to tackle issues of common concern.

Science in the Arab region

The unfortunate reality is that the state of science and technology in the Arab region, which was comparable to that in Brazil, China, India and Mexico three decades ago, has fallen substantially behind these countries in the ensuing years.

Individual countries have primary responsibility for improving the state of science within their borders. Yet organizations like TWAS can play a prominent role in this effort, as the Academy has shown time and again over the course of its history. In fact, the progress that has been achieved – often with the help of TWAS – in a growing number of countries now allows the Academy to focus attention on regions and countries that still need assistance.

To advance its goals, TWAS should continue to recognize and honour scientific achievement, placing a spotlight on individuals and institutions that pursue scientific excellence. With other organizations such as UNESCO and IAP, the global network for science academies, it should work to create science academies in countries where they do not exist and help to strengthen science academies in countries where they do. TWAS should also encourage regional cooperation in science, serving as a bridge between scientific institutions both in the South and the North. And it should use its presence and credibility throughout the developing world to function as an influential advocate for science.

All of this should be done hand-in-hand with the Academy's partners, especially those countries in the South that have attained unprecedented progress, thanks in large measure to investments in science and technology. These countries not only have the resources to help but also now serve as sterling examples of what can be accomplished in a brief period.

The Arab Spring began as a quest for democracy, dignity and economic security. These are principles that animated the majority of people across the region, especially the young, university-educated segments of the population who, in many ways, represent the region's future. Subsequent events, most notably the violence and repression that has afflicted a number of countries, have clouded the region's future. No country, even those that have remained relatively peaceful, can escape what is happening. The region is too small and too intertwined for any country to be sheltered from the ferment that surrounds it.

The Arab Spring may have turned into the Arab Autumn, but hope for a better future remains. The scientific community – both locally and internationally – cannot change the course of history on its own. Science itself cannot hope to advance in the midst of violence and chaos.

Nevertheless the scientific community, with its emphasis on transparency, .merit and excellence, can help steer events in a positive direction if science is allowed to take hold within society. By helping science succeed in regions such as mine, TWAS can play as vital a role today as it has in the past. In this way, the Academy's 30th anniversary milestones can be a prelude to even greater success in the future.

Full engagement

Zakri Abdul Hamid

Zakri Abdul Hamid, who was born in 1948 in Bentong, Pahang, Malaysia, 80 kilometres northeast of Kuala Lumpur, has enjoyed a distinguished career as an educator, researcher, administrator and diplomat. He has been a leading player in Malaysia's efforts to build scientific capacity as a key element for promoting sustainable economic growth. He has also been a prominent figure in international organizations dedicated to the conservation and sustainable use of biodiversity and ecosystem services.

In 2010, Zakri was appointed science advisor to the prime minister of Malaysia. He currently serves as the government's chief spokesman on scientific issues and chairs or co-chairs the National Science and Research Council, the Malaysian Biotechnology Corporation, the Joint Malaysian Industry-Government Group for High Technology (MIGHT) and the Aerospace Malaysia Innovation Centre. In addition, he is co-secretary of the Global Science and Innovation Advisory Council for Malaysia, which is chaired by the prime minister. The Council is a joint initiative of MIGHT and the New York Academy of Sciences in the United States.

In January 2013, Zakri was elected founding chair of the Intergovernmental Science-Policy Platform on Biodiversity and Ecosystem Services (IPBES). IPBES is open to all UN member states. Its mandate is to assess the state of global biodiversity and ecosystem services and to devise strategies for protecting these invaluable but threatened resources.

In September 2013, Zakri was appointed to the UN Secretary-General's Scientific Advisory Board, established by Secretary-General Ban Ki-moon to provide high-level advice on the role of science, technology and innovation for sustainable development.

Zakri served as director of the United Nations University's Institute of Advanced Studies (UNU-IAS) from 2000 to 2008; co-chair of the Millennium Ecosystem Assessment Board from 2000 to 2005; secretary-general of the Society for the Advancement of Breeding Research in Asia and Oceania (SABRAO) from 1981 to 1989; and chairperson of the Subsidiary Body on Scientific, Technical and Technological Advice of the Convention on Biological Diversity (CBD) from 1997 to 1999. He was founding president of the Genetics Society of Malaysia and deputy vice-chancellor of the National University of Malaysia from 1992 to 2000.

He received a diploma from the College of Agriculture in Malaya, Serdang, Malaysia, in 1969, a bachelor's degree from Louisiana State University in 1972, and master's and doctorate degrees from Michigan State University in the United States in 1974 and 1976. His major research interests focus on the breeding of rice and soybean in Malaysia and genetic variations of indigenous Malaysian timber species. He has written and contributed to numerous books and articles both on these subjects and on biodiversity and biotechnology policies in developing countries.

He has been the recipient of many honours, including the Fulbright Fellowship, Rotary Research Foundation Gold Medal, Langkawi (Environment) Award and Zayed International Prize for the Environment. He holds fellowships in the Academy of Sciences Malaysia, World Academy of Art and Science and the Islamic World Academy of Science, and has received honorary doctorates from Universiti Malaysia Terengganu, Universiti Sultan Mizan, *University of Nottingham and* Universiti Putra Malaysia.

In a rare distinction, three species have been named after him: Paleosepharia zakrii *(a beetle);* Pomponia zakrii *(a cicada); and* Nephenthes zakriana *(a carnivorous pitcher plant, whose morphology is characterized by liquid-filled cavities designed to lure and capture their prey).*

Zakri has pursued a career dedicated to bringing scientific researchers and public officials closer together as part of a larger endeavour to address a critical global challenge: the loss of biodiversity and the ecosystem services that biodiversity provides. In the following essay, he speaks about his efforts to advance this goal and the role that international organizations such as TWAS have played – and should play – in this effort.

***

I cannot recall a particular moment in time when I first learned about TWAS. I must have heard about it soon after it was launched in 1983 as news about a new international science academy dedicated to the concerns of scientists in the developing world circulated in Malaysia's scientific community. Abdus Salam, TWAS's founding president, was a well-known figure among scientists in Malaysia. Any significant measures he took to promote science in the developing world would likely have been discussed in my home country.

The Malaysian scientific community, of course, was not alone in this regard. Salam had the capacity to command the utmost respect and attention from scientists across the developing world. His accomplishments as a Nobel laureate and one of the world's preeminent physicists made him a singular figure in science in the South. His relentless efforts to promote science in developing countries only enhanced his reputation and acclaim. There has been no one else like him – then or now.

My generation

The global activities that Salam engaged in, including the creation of TWAS, spoke to the needs and aspirations of my generation of scientists who were well trained, idealistic and eager to make a mark on the world – increasingly so by contributing to their own country's efforts to build scientific capacity. I think that it is fair to say that rarely have the man and the moment come together so seamlessly as they did in the case of Salam.

In the early 1980s, I was a young associate professor in the Department of Biology at the *Universiti Kebangsaan Malaysia* (the National University of Malaysia). At the time of TWAS's launch, I had just been appointed head of the Department of Genetics. Several years earlier, I had returned to Malaysia from the United States after earning advanced degrees at Louisiana State University and Michigan State University.

I did not feel a sense of isolation – at least not to the extent that Salam had felt upon returning to Pakistan in the early 1950s after earning a doctorate degree in physics at Cambridge University in the United Kingdom. I never doubted that I would come back home after receiving my degrees.

As an added incentive, I had a contractual obligation to do so. The *Universiti Kebangsaan Malaysia* had awarded me a full scholarship to travel to the United States for graduate studies. The award carried the stipulation that I would return home after obtaining my degrees.

I could, however, likely have broken the contract without any ramifications. So, it wasn't the legalities that brought me back. Rather, it was a deep commitment to contribute to the welfare of my country – a commitment that I shared with many other young researchers in Malaysia.

I thought that my area of study – plant breeding and genetics – had vast applications for improving Malaysian agriculture and conserving and utilizing its treasure trove of natural resources. Such efforts, I believed, could pay huge dividends in enhancing Malaysia's economic, social and environmental well-being. At a personal level, I was convinced that these national challenges offered a pathway for a successful and rewarding career.

In sum, as a young, well-trained researcher with strong ties to Malaysia's scientific community, I firmly believed that I could utilize my education to make a difference in my country.

Nevertheless, I did not feel completely at ease with either my research or career prospects when considering my interest in collaborating with the international scientific community. In short, at a national level, I did not feel isolated. At an international level, I did.

Beyond borders

It was a desire to be fully engaged with the global scientific community that made me enthusiastic about TWAS. I believed that the organization could be a valuable forum for helping scientists in the developing world reach beyond their national borders and engage with colleagues across the globe, especially in the South.

This has been – and continues to be – one of the great contributions of the Academy to both science and society. In my mind, it is one of the primary factors that make TWAS as relevant today as it has been since its inception. Among scientists, the need for international cooperation is always a necessity and never falls out of fashion. Science, as Salam said on so many occasions, is a global enterprise that knows no boundaries.

As I mentioned previously, I was not directly involved in the Academy during the early years of my career. I was, instead, introduced to TWAS indirectly by some of its most eminent members.

For example, in the early 1980s I met TWAS Founding Fellow M.S. Swaminathan, the world-renowned agricultural scientist and a chief architect of India's Green Revolution, at a conference focusing on improving rice yields in the developing world. Swaminathan was then the director general of the International Rice Research Institute (IRRI).

It was there that I also met Gurdev S. Khush (TWAS Fellow 1989), a distinguished plant geneticist and long-time researcher and principal plant breeder at IRRI. Khush has won virtually every international prize for agricultural research, including the Japan Prize, the President of India's Padma Shri Award, the Wolf Prize and the World Food Prize.

Research collaborations soon followed with both of these renowned scientists.

Because they were members of TWAS, by extension the Academy slowly became part of my community – or, perhaps more accurately, my science fraternity – since my early association was largely guided through individual members and not the organization itself. It was Khush who nominated me for membership in TWAS, resulting in my election in 1996.

By then, my career had become firmly established. At the time of my election, I was serving as deputy vice-chancellor at the *Universiti Kebangsaan Malaysia*. I had also become increasingly involved in both national and international organizations dedicated to building scientific capacity and strengthening science policies in both Malaysia and throughout the developing world, especially in the field of biodiversity and plant genetics. In my own country, I had chaired a task force directed to prepare the National Policy on Biological Diversity and was the founding president of the Genetics Society of Malaysia.

On the international front, I was a Malaysian delegate to the negotiations on the UN Convention on Biological Diversity and later headed the delegation to the meetings of the treaty's Conference on Parties. I also had led Malaysia's delegation during the initial round of negotiations of the Cartagena Protocol on Biodiversity. At the regional level, I had served as secretary-general of the Society for the Advancement of Breeding Research in Asia and Oceania (SABRAO).

I think my election to TWAS was based on both my research and my involvement in advancing scientific capacity and sustainable science-based development in my own country and, more generally, in developing countries, often through international organizations.

Broader reach, higher profile

Soon after my election, Mohamed Hassan, the Academy's executive director, appointed me to several TWAS committees that were seeking to strengthen the Academy's ties to other organizations, especially within the UN system. At the time, international organizations were expanding their efforts to promote science and technology, especially for the purposes of poverty alleviation and economic development.

TWAS, with a membership list that included many of the developing world's most eminent scientists, was poised to serve as an important ally in broader global efforts to address the South's most critical challenges, including efforts to promote sustainable economic growth.

In the late 1990s, for example, TWAS reached out to UNESCO, to extend its relationship beyond UNESCO's responsibilities for the Academy's finances and personnel to joint programmatic activities; to the UNDP, to devise collaborative projects on issues of common concern, especially examples of successful applications of science and technology to promote economic development; and to the Global Environmental Facility (GEF), to explore best practices in science-based development in the developing world with the help of GEF funding. TWAS was taking steps not only to gain a greater presence within the international science community but also within the international economic development community.

The Academy did make some progress in forging greater and stronger partnerships. For instance, it emerged as a strong and credible voice for Southern concerns and interests during the organization of international conferences sponsored by UNESCO and the International Council for Science (ICSU). This included preparations for the World Science Conference in 1999, which was held in Budapest. At the same time, it was able to expand joint fellowship programmes with, for example, UNESCO and ICSU, and also to pursue a GEF-funded project examining issues related to dry lands in the developing world. It also partnered with the UNDP Special Unit for South-South Cooperation to produce a series of booklets showcasing examples of successful applications of science and technology to meet critical challenges in the South.

Progress in forging closer collaborations with other international organizations was slower than we might have liked. Nevertheless progress was made. The result was that the Academy was able to broaden its reach and raise its profile in the international and economic development communities.

A matter of equity

The late 1990s and early 2000s were a critical time for TWAS. As the level and nature of its partnerships expanded, so too did its global reputation. In a sense, this period marked the arrival of TWAS on the world stage – a position it has held and amplified ever since, not only to the benefit of the Academy but also to the benefit of science and, increasingly, society.

I was first elected to the TWAS Council in 2001 and remained a member for the next decade. My engagement with the Academy at this time included a period as vice-president for the Central and South Asia region from 2004 to 2006. It was then that I became more involved with the internal challenges that the Academy faced.

For example, the Council was concerned that the Academy had become dominated by the developing world's largest and most populous countries – notably, Brazil, China and India – and that there was a need to broaden representation from other countries that had yet to reach comparable levels of scientific capacity (and, in fact, that in some instances had actually fallen farther behind).

Demographically, small and sparsely populated countries tended to be under-represented in TWAS – or not represented at all. Geographically, Africa and countries with predominantly Muslim populations also tended to be under-represented not only among TWAS members but also among the Academy's fellowship awardees and prizewinners.

The issue was a disconcerting one that struck at the heart of two of the Academy's fundamental principles: excellence and equity. On the one hand, scientists were elected to TWAS because of their accomplishments – that is, the excellence of their work. On the other hand, the Academy was dedicated to building scientific capacity in all developing countries, which meant that it had an obligation to pursue activities and policies that would boost access and equity within the Academy among countries that had continued to lag behind in their scientific capacity.

Stated another way, the Council believed that it would be a serious mistake for the Academy to compromise its emphasis on excellence. That would weaken one of the underpinnings of its success and ultimately compromise its credibility. Yet the Council also believed that it would be a serious mistake to ignore the growing imbalances in its membership, where just a handful of countries dominated the Academy and largely set the agenda. The latter trend, if it became irreversible, would help create a South-South gap in science and technology potentially as damaging to global science as the North-South gap had been.

Except for creating a special grant fund for collaborative research projects among Least Developed Countries (LDCs), the Council chose not to make any fundamental changes in the way it elected its members or distributed its fellowships, awards and prizes. Instead, it decided to redouble its efforts to find talented scientists in countries that were underrepresented in the Academy, primarily by relying on its members and its regional offices to tap into their networks to find colleagues worthy of recognition but who had not yet been recognized largely because of where they lived.

As a result, the response to the challenge of geographic imbalance within the Academy was largely informal. This meant that progress was slow. But it was a strategy that retained an abiding principle that had been enshrined in the Academy by Salam and the founding members: First and foremost, TWAS would acknowledge and reward scientific excellence.

Gender balance

A second critical challenge that the Council grappled with in the late 1990s and early 2000s was the small number of female members. This is a problem that has beset not only TWAS, but all science academies. Here, again, the Council sought to balance its emphasis on excellence while serving as a source of opportunity for an under-represented group.

TWAS established no special programmes to address this issue. Instead it took steps to identify worthy female candidates for membership and its various fellowships and grants. The Academy also continued to support the Third World Organization for Women in Science (TWOWS), an organization that TWAS launched in 1988 with the specific mandate to advance the interests of female scientists in the developing world.

The Academy's efforts on behalf of TWOWS were, in a larger sense, a form of capacity building. I am delighted to see that the organization – now called the Organization for Women in Science for the Developing World (OWSD) – has secured independent funding for a broad range of activities designed to promote women scientists both in the South and globally.

The growing impact of OWSD reflects the growing strength of women within the scientific community and society overall. The benefits generated by organizations like OWSD not only manifest themselves in the breadth and depth of the international scientific agenda but also in challenges related to poverty reduction, environmental protection and sustainable development.

Serving youth

A third critical challenge facing TWAS in the late 1990s and early 2000s was the need to address the concerns of young scientists and, more specifically, to devise strategies that encouraged young people to pursue degrees and careers in science.

Like the challenges posed by increasing the presence of women in the Academy, the challenges presented by efforts to engage young scientists are a global phenomenon not relegated to the developing world. Yet in the South it has been an even more crucial issue than in the North.

That is because any progress in scientific capacity that has been achieved in the South could be quickly reversed if subsequent generations of scientists were to be too few in number or inadequately trained. A single generation of well-educated scientists clearly would not be sufficient for countries that have just begun to build their scientific capacity. As the first generation of scientists retired, there would be an insufficient number of scientists in the wings to take their place.

This phenomenon surfaced in African countries in the 1980s and 1990s. Kenya, Nigeria and Sudan, for example, had recruited scientists to work in their emerging universities and research centres in the 1960s and 1970s only to experience a shortage of scientists 25 years later as the first-generation scientists retired and were not replaced, due, in part, to the absence of scientists in their 30s and 40s. Unlike the North, these countries could not compensate for this loss by attracting scientists from beyond their borders. They were, in fact, the victims, not the beneficiaries, of brain drain.

When it came to helping young scientists, the Council not only pursued informal measures (for example, asking members to identify promising young scientists in their countries and regions) but also took concrete steps to address this issue.

For example, TWAS created a Young Affiliates programme based on a new category of (temporary) membership for scientists less than 40 years of age. It also sponsored awards programmes for young scientists, both on its own and in partnership with other institutions, including the African Union. And it encouraged interactions and networking among young scientists by supporting workshops and conferences.

The Academy's regional offices, with support from the TWAS secretariat, organized these activities. Such efforts represented a new front in the Academy's campaign to build scientific capacity by aiding the next generation of scientists in the South.

A bridge between cultures

In 2000, I was named co-chair of the Board of the Millennium Ecosystem Assessment (MA) and director of the United Nations University's Institute of Advanced Studies (UNU-IAS). These appointments shifted my career focus from teaching, research and university administration to one engaged primarily in international science.

My responsibilities focused on drawing the scientific and policy communities closer together to enable science to play a more prominent role in addressing the world's most critical issues – for example, poverty alleviation and sustainable economic development; food, energy and water security; natural resource degradation; biodiversity loss; and protection of the ecological services that biodiversity provides.

MA provided the first comprehensive global appraisal of the recent unprecedented loss of biodiversity and outlined the consequences that such an ominous trend would have on the quality of life for citizens in both the North and South.

The assessment noted that developing countries would likely pay the highest price for biodiversity loss because people living in poor countries relied on the Earth's natural resources for their livelihoods and well-being to a greater extent than people living in developed countries. Moreover, in many instances, they did not have the financial means to mitigate the adverse impacts that accompanied the loss of these resources.

In my work with the UNU-IAS, we sought to expand the organization's responsibilities as a think tank for the UN system that would analyse a broad range of issues – energy, climate change, desertification, infectious diseases, transportation, urbanization (the list was extensive).

The institute would also seek to build capacity not just in science but also in science policy, serving as a bridge between researchers and public officials. Conferences and training sessions, particularly for young professionals pursuing careers as policy analysts and administrators, were the primary mechanisms by which UNU-IAS hoped to advance this goal.

In all my efforts with the MA and UNU-IAS, I felt a strong affinity with TWAS. I believed that the missions of the organizations concurred with those of the Academy. This was especially the case as TWAS became increasingly engaged in policy issues.

Historically, it has been the nature of science academies – and, more generally, the scientific community – to shy away from the political arena and to avoid getting their hands dirty in the often raucous world of politics.

I fully appreciate the concerns that scientists and scientific institutions have when they move beyond their areas of expertise. And I am by no means suggesting that individual scientists or scientific institutions should ignore the risks. If either individuals or institutions become too closely associated with a political organization – or even a particular political viewpoint – it could jeopardize their standing as impartial players in the public policy arena. Even more importantly, funding could be jeopardized if political fortunes shift, as they often do.

However, I do think that the world needs – and is increasingly soliciting – the expertise that scientists and scientific institutions have to offer. At the same time, academies in both the North and South have become increasingly comfortable – indeed eager – to offer their expertise to address critical societal concerns.

A bigger force in science policy

TWAS is well positioned to take advantage of the growing interest that the public and policymakers have in seeking out the scientific community for advice. There are, moreover, models that TWAS can turn to – for example, the US National Academy, the Royal Society and the Chinese Academy of Sciences – when trying to hone its skills as a source of information and insight on critical policy issues. The Academy has the credibility, expertise and, I believe, the organizational interest and skills to move ahead on this front.

TWAS's modest budget, combined with the enormous scope and complexity of the challenges, would likely lead to modest results – at least at first. The Academy is an international organization. Consequently, it would likely render its greatest value by focusing on broadly cast global issues. TWAS's limited resources, however, would not even begin to match the breadth of the challenges it would be seeking to address. Despite this, I believe the Academy should expand its efforts to become a larger player in the science policy community. In my view, this is TWAS's next great opportunity for acquiring a bigger platform for societal influence and impact. Over time, patience and persistence could lead to impressive results.

I have been thinking about these issues a great deal ever since my appointment as the science adviser to the prime minister of Malaysia in 2010.

My country envisions becoming a high-income country by 2020. The goal is within our grasp. But it can only be reached if science is fully integrated into the nation's social and political fabric. This means research and development must be a priority; adequate investments in human resources must be made; science education must be improved, not just for students with aptitudes in science but for all students; efforts must be made to foster stronger public appreciation for science; and measures must be taken to ensure an adequate percentage of a nation's gross domestic product (GDP) is devoted to research and development (R&D).

The historical record shows that 2.5% of GDP is a minimum share of the nation's wealth that should be invested in R&D. In Malaysia, the figure currently stands at just 1.07%. The trajectory has been in the right direction but we obviously have some way to go before we can claim that the level of our outlays are comparable to those of other nations with strong scientific capabilities.

TWAS could play a vital role in encouraging not only developing countries but all countries to pursue aggressive strategies for scientific capacity building. It would be engaging in such efforts from a global perspective that would make it unique among scientific academies and likely enhance its effectiveness.

Whether the issue is science education and training; public awareness of science; scientific infrastructure; water, food and energy security; public health; or virtually any other critical concern facing the world today, the voice of science – as articulated by institutions like TWAS – is essential for informed decision-making and, hopefully, productive outcomes.

That is why I would like to see the Academy focus a good deal of its attention on promoting closer connections between the policy and science communities both in the South and globally.

The effort would require a willingness by scientists to emerge from their sheltered universities and academies and to appreciate the responsibilities and pressures faced by public officials. At the same time, it would require a willingness by public officials to adopt evidence-based policies regardless of how difficult such efforts may be in the short-term.

For many of the world's challenges, we now have sufficient knowledge to pursue more effective policies than we have in the past. This is due in some measure to the research that has been done over the past several decades by scientists in the developing world.

I am not suggesting that research agendas should be narrowed or that research activities, especially in the basic sciences, should be curtailed. But I am suggesting that we do not need to wait for additional information to act.

Lost biodiversity

A fundamental challenge for the scientific community is to present its knowledge in ways that can be readily understood by policymakers and the public and that provide potential pathways to address the increasingly complex challenges faced by nations on their own and by the global community collectively.

That is the agenda the Intergovernmental Science-Policy Platform on Biodiversity and Ecosystem Services (IPBES), a newly established UN organization, for which I am privileged to serve as the founding chair, plans to pursue in the years ahead.

The IPBES was launched in early 2013 after nearly a decade of discussion. It will provide the largest assessment of the global ecosystem in history. More than 2,000 experts from 100 countries are expected to participate.

Our objective is to place biodiversity loss and the threat this loss presents to vital ecosystem services high on the agenda of international concerns. The risks posed by biodiversity loss are comparable, in some ways, to the risks posed by climate change. In fact, to some extent, the two are related.

Like climate change, concerns for the loss of biodiversity and the damage incurred to ecosystem services are difficult for the public to appreciate and respond to because the changes are long-term, problems often surface at a slow pace, solutions are not always straightforward, and effective responses, particularly at the individual or even community level, are difficult to fathom.

That's where scientists and scientific institutions can play a leading role by explaining the nature of the problem and what can be done about it in a language that everyone – and particularly public officials – can understand.

That's what IPBES hopes to accomplish: to present to as large a number of people as possible critical trends in biodiversity loss, what it means to individuals, nations and the global community, and how we can address the challenge together. We welcome the opportunity to forge strong ties with TWAS and other academies in our efforts.

If science shows the way

If Abdus Salam were alive today, he would undoubtedly be proud of what the Academy has accomplished over the past 30 years. Yet he would also be the first to say that the job is not yet done. He would, moreover, remind all those who would listen that the ultimate goal is not scientific capacity building but improving the lives of all people – a goal that can only be reached if science shows the way.

Showing the way has been the hallmark of TWAS in the past, and I am confident it will remain a hallmark of the Academy in the future.

My generation of scientists in the developing world is indebted to TWAS for what it has meant to our careers. It is a debt that we can repay by continuing to contribute our knowledge and expertise to society's efforts to address critical challenges and by supporting the Academy in its efforts to advance its worthy goals.

Meeting & 25th Anniversary Celebration

The odds of success

Keto Mshigeni

Keto Mshigeni, one of Africa's preeminent natural scientists, was born in 1944 at Mpinji, Mamba, a rural village in northern Tanzania. He received his primary education in his hometown and his secondary education at the Tabora School for boys. He earned a Bachelor of Science degree from the University of East Africa (UEA), Dar es Salaam University College, in Tanzania in 1969, and a PhD in botanical sciences from the University of Hawaii in the United States in 1974.

Throughout his career, Mshigeni has embraced a deep commitment to transforming scientific knowledge into effective strategies for income growth and economic development, especially at the community level. His goal has been to help improve the economic and social well-being of Tanzania's – and, more broadly, Africa's – poorest people.

Mshigeni is widely recognized for his research on seaweeds and mushrooms. He has helped to turn these wild plants into commercially farmed crops that are grown on a small, yet profitable, scale in Tanzania and other African countries. During his career, he has also acquired an encyclopaedic knowledge of Africa's ecology and its trove of natural resources. He has put this knowledge to work in promoting sustainable development across the continent and especially in Tanzania.

Mshigeni has published more than 150 scientific articles. In addition, he has written numerous academic books, pamphlets, and training manuals for scientists and the public, as well as articles and editorials that explore the wonders of science and the environment for a broad audience. He served as editor-in-chief of Discovery and Innovation, *the flagship publication of the African Academy of Sciences (AAS) from 1994 to 2010, and currently sits on the editorial board of the* African Journal of Ecology *and the* International Journal of Phycology and Phycochemistry.

In his writings, Mshigeni has often sought to instil a sense of appreciation and curiosity for science among young people in Africa. This sentiment has been most recently expressed in a book, Lighting a Fire: Inspiring Stories of 31 Eminent Tanzanian Scientists, *co-edited with Ludovick Kinabo and published by the Tanzania Academy of Sciences.*

Mshigeni is currently vice-chancellor at the Hubert Kairuki Memorial University. He previously served as pro-vice-chancellor for academics and research at the University of Namibia, and professor of botany and director of postgraduate studies at the University of Dar es Salaam in Tanzania. In the mid-1990s and early 2000s, he was the United Nations Educational, Scientific and Cultural Organization/United Nations University (UNESCO/UNU) Africa chair for ZERI (Zero Emissions Research Initiative). During the same period, he directed the United Nations Development Programme (UNDP) Regional Project on Sustainable Development from Africa's Biodiversity.

Mshigeni is a recipient of the AAS/Ciba-Geigy Prize for Agricultural Biosciences, the Boutros Boutros-Ghali Prize for the promotion of science in Africa, and the Association of African Universities Award of Excellence in Higher Education and Research in Africa. He is a fellow of AAS, the Tanzania Academy of Sciences and the World Technology Network.

In 2012, Tanzania President Jakaya Kikwete appointed him chairman of the Board of the Tanzania Atomic Energy Commission and, in August 2013, he was elected vice-president of the International Medicinal Mushrooms Association. Few other scientists in Africa – or elsewhere – can lay claim to such a wide range of interests and expertise.

In the following essay, Mshigeni describes how he has devoted his scientific career to a broad social purpose becoming, in effect, a scientist for the people. He also examines TWAS's role in helping to build scientific capacity in sub-Saharan Africa. In the concluding pages, he outlines additional steps that he thinks the Academy should take to strengthen its impact on science and society across the continent.

***

I first encountered Abdus Salam in March 1984 when I was a professor of botany and director of Postgraduate Studies at the University of Dar es Salaam in Tanzania. Mwalimu Julius Nyerere, founding president of Tanzania, had invited Salam to speak at the university.

Nyerere and Salam were colleagues – indeed friends – who shared a deep concern for the dismal economic and social conditions in the developing world. Both were determined to chart an effective course for improving the material prosperity of impoverished people throughout the South and both believed that science would play a critical role in these efforts.

In 1987, President Nyerere, in his capacity as chairman, appointed Salam to the South Commission, a path-breaking international agency dedicated to examining development issues in the South. As a physicist, Salam held a unique position in an organization that had shown a greater affinity for economists and social scientists than it had for natural scientists. The Commission's interest in devising a pathway for poverty alleviation and sustainable development has been subsequently pursued by the South Centre, an intergovernmental organization headquartered in Geneva, Switzerland, which proudly proclaims to have been created "by the South for the South".

In line with the organization's mission, the Commission's historic 1990 report, *The Challenge to the South*, examined the problems of development primarily through the eyes of the South's political and intellectual leaders.

Equally important, and thanks largely to Salam (and Nyerere's unwavering support for Salam's ideas), the report also emphasized science and technology's central role in improving the economic and social well-being of developing countries. Not surprisingly, it focused a great deal of attention on the importance of South-South cooperation in science in advancing its goals.

The Challenge to the South (notice the use of the word "to" and not "of") was an extraordinary document conveying a message that still resonates today. While the report did not dismiss the value of assistance from the North, it made clear that the South would have to assume responsibility for its future and that acquiring the requisite knowledge and skills, especially in science and technology, would be the keys to progress.

For the sake of the people

I did not know Salam when I heard him speak for the first time at the University of Dar es Salaam in 1984. His name was unfamiliar to me despite his fame and global status as a Nobel laureate. My major areas of research, botany and ecology, resided on the other side of the science spectrum, far afield from physics and mathematics.

President Nyerere, however, may have mentioned my name to Salam, knowing that Salam was always eager to learn about scientists in the developing world, especially those pursuing cutting-edge research that was having a positive impact on the lives of people in the South.

At the time of Salam's visit, I was leading efforts to build seaweed demonstration farms on the Tanzanian coast. In 1983, I had published a booklet in Kiswahili, *Mwani: ukulima wake baharini na manufaa yake kwetu* ("Seaweeds: their cultivation in the sea, and their benefits to us") to help local citizens learn productive seaweed farming practices. The goal was to promote seaweed farming in Tanzania's coastal waters by providing villagers with an understanding of biology, ecology and growth dynamics.

Salam was interested in hearing about scientists like me, not just as a matter of curiosity, which he had in abundance, but also as part of a larger desire to put us to work on his broader agenda for poverty reduction and science-based economic development. One of his goals was to assemble a "science corps" across the developing world dedicated to improving the conditions of poor people in the South. When Salam spoke of building scientific capacity in the developing world, it was not simply for the sake of science. Ultimately, it was also for the sake of the people. TWAS was created, in part, with this lofty vision in mind.

The research I was doing on seaweed, especially my efforts to commercialize it in Tanzania on a small, community-based scale, had attracted the attention of both the scientific and popular press. The media hailed my efforts as an example of how science could help lift the local economy and lessen poverty among the most vulnerable segments of the population.

President Nyerere and those close to him often cited our efforts as a model for how scientific knowledge could serve as a critical tool for development. I believe that Tanzanian officials were keen to inform Salam about my research and outreach initiatives. I also believe that Salam asked officials at Dar es Salaam to give him a copy of my *curriculum vitae*. The information contained in my CV became the basis of my election to TWAS in 1987.

Unifying force

Consequently, my unfamiliarity with Salam, which still surprises me when I think about it, did not stand in the way of getting to know him. Nor did it impede my ability to participate in the worthy organizations – most notably, TWAS – that he had created. I thank all those who worked on my behalf – often without my knowledge – to make this happen.

Along the way, I served as chairman of the TWAS Agricultural Sciences Committee. I was also accorded the honour to deliver a TWAS Medal Lecture at the TWAS General Meeting in Mexico City in 2008. And I was chosen to serve on the TWAS Council and as vice-president of the Academy. It has been a thrilling experience, especially considering my humble beginnings.

Among all his other remarkable qualities, Salam was a gifted orator. It was not so much his delivery, marked by his stately English accent, which accounted for his allure as a public speaker. Rather, it was the mesmerizing power of his ideas – his ability to examine complex concepts in clear and convincing ways.

When Salam spoke, he commanded the room like few other scientists could. As a renowned researcher and Nobel laureate, as a learned intellectual conversant in the writings of the East and the West, and as a committed reformer who had gained enduring respect both within and beyond the global scientific community, he was a compelling presence on the podium. When he spoke, image and content became a unified force of persuasion.

Some three decades have elapsed since I first heard Salam speak at the University of Dar es Salaam. I have subsequently listened to hundreds, if not thousands, of talks. But Salam's lecture was different. I still recall what he had to say. The themes of his talk spoke directly to my own efforts to define my work and career, even though I was already a professor.

His words tapped a deep sentiment within my personal reveries that I had yet to express in a coherent and convincing manner, either to myself or to others.

Salam spoke eloquently about knowledge as a unifying force for humankind. He also emphasized that the pursuit of knowledge was a continuum.

In a sense, he was claiming that neither knowledge nor those who pursued it could operate solely on their own. Quite the contrary, Salam maintained that knowledge was a collective force that transcended national and disciplinary boundaries. He stressed that scholars and scientists depended deeply on the efforts of others, both past and present, in their quest to unlock the mysteries of the universe and to speak to universal values and truths. As Salam often noted, knowledge – and particularly scientific knowledge – was "the common heritage of all humankind".

The sense of unity that Salam conveyed in his speech closely tracked my beliefs that science is an instrument for good among all people. The four-fifths of humanity who did not have adequate access to scientific knowledge – and therefore could not turn to science to address their basic needs – were burdened by a fundamental handicap that largely accounted for their poverty and marginalization. From Salam's perspective, scientific knowledge was more important than ever. Countries that failed to acquire it were consigning their people to lives of poverty.

Noble pursuits

Salam spoke of science as a noble pursuit in ways that confirmed the value of my own work. Similarly, his discussion of science as a continuum not only reinforced the growing importance of interdisciplinary research, but also suggested that singular scientific findings often gleamed far beyond the original insights that had been brought to light.

From a practical viewpoint, this meant that my studies of the *Eucheuma* seaweed likely carried implications beyond that particular plant. (I prefer to use the word plant, not weed.) This was something I would subsequently discover as I expanded my research agenda to include mushrooms and other native plants that grew in the wild but had yet to be successfully cultivated in Africa.

Salam, moreover, seemed to contend that broad principles of science, whether in physics, biology or other disciplines, could be applied both within and beyond science for the benefit of the larger society. In Salam's mind, science's universality was not only due to its findings but also to its methodologies (its emphasis on transparency, its insistence on replicability and its devotion to excellence). These core principles drove Salam's efforts both as a scientist and a humanitarian.

On a more personal note, thanks in part to my encounters with Salam, the same principles that he outlined in his lecture at the University of Dar es Salaam helped to propel my quest for a productive and meaningful career as a scientist of and for the South.

Striving to reach the villages

In retrospect, it seems as if my career and Salam's lecture intersected at an opportune moment. My research, at the time, was having a modest but discernible impact on Tanzania's agricultural economy. Women, in particular, were eager to learn and apply the non-traditional agricultural practices that were at the heart of my training efforts. They viewed the training as a welcome opportunity to increase their meagre incomes.

These efforts, in turn, raised my profile in Tanzania's scientific and economic development communities, improving my prospects for a secure and productive career in science in my home country, which was something that I desperately wanted to do.

Salam was not the only prominent person to influence my career. President Nyerere was also a great inspiration. The trajectory of my career, in fact, was influenced by sentiments that he conveyed on 25 October 1961, when speaking at the inaugural ceremony of the Dar es Salaam University College. He presented this challenge to the faculty: "While other people can aim at reaching the moon, and while in the future we might aim at reaching the moon, our present plans must be directed at reaching the villages."

President Nyerere wanted our university to be directly involved in uplifting our impoverished rural people. This message became the guiding principle throughout my university career, well before hearing Salam speak for the first time.

In the 1960s, Nyerere was not only the president of Tanzania but also the chancellor of the University of East Africa (UEA). The consortium-like university, with initial ties to the University of London in the United Kingdom, served East Africa's Great Lakes region. Engineering was the focal point of study at Nairobi University College in Kenya; law at Dar es Salaam University College in Tanzania; and medicine at Makerere University College in Uganda. In 1970, UEA split into three independent universities: the University of Nairobi, Makerere University and the University of Dar es Salaam.

In much the same way that Salam's goals were shaped by his personal experience, my response to President Nyerere's call was due, in large measure, to my life's journey. I was born into a family of modest means and grew up at Mpinji, Mamba, a small village in Tanzania's Kilimanjaro region.

My earliest years of formal education took place in a local school in my village. My secondary schooling took place at the Tabora School for boys. Created as a centre of learning for the sons of tribal chiefs during the colonial era, the school was subsequently opened to children of ordinary parents like mine because the children of chiefs alone could not fill the classrooms. Admission, in my time, was based solely on scholastic ability. Only the country's most academically gifted boys were admitted.

Upon graduating from the Tabora School, I was admitted to Dar es Salaam University College, where I earned an undergraduate degree in botany, geography and education in 1969.

It was in a botany laboratory in my second year at Dar es Salaam College when I first became interested in seaweeds.

I hailed from the mountains, a far distance from the sea. My initial encounter with seaweeds was through the study of plants in biology textbooks. When I saw, for the first time, samples of seaweeds freshly collected from marine waters near Dar es Salaam, I found their diverse shapes, sizes and colours to be endlessly fascinating. In lectures and discussions since then, I have often referred to my seaweeds infatuation as "love at first sight".

Erik Jaasund, a professor of botany who had been recruited from Tromsø University in Norway to teach at Dar es Salaam University College campus and who had shown me freshly collected seaweed plants for the first time, was responsible for stirring my interest.

Jaasund was an inspiring scholar and an enthusiastic teacher. Two weeks after he had seen me intensely examining samples in his laboratory, he invited me on a weekend trip to the Tanzanian coast to observe seaweeds in their natural habitats. This adventure marked the first significant step in a journey that has guided much of my research agenda for more than four decades now.

An accident of history

I have often thought of my career as an accident of history. Of the more than three billion people in the world at that time, I was fortunate to discover someone who appreciated seaweeds as much as I did; who had acquired a great deal of knowledge about these unusual marine plants; and who was eager to share this knowledge with others – me! If this isn't serendipity, I don't know what is.

Jaasund kindly paved the way for me to secure a small grant from the Norwegian Agency for Development Cooperation (NORAD) that allowed me to work under his guidance from April 1969 to March 1970. During this year, I took numerous field trips with him, exploring the full length of the Tanzanian coast. I gathered hundreds of samples of seaweeds and compiled voluminous notes on their habitats. I also spent a good deal of time reading about seaweeds.

My research revealed that humans had used seaweeds since ancient times. But it also showed that seaweed farming was a recent phenomenon, first taking place in China and Japan in the 17th century. In Tanzania, seaweeds had been harvested in the wild over aeons of time and used, among other things, for baiting fish, dressing wounds and treating rashes and skin diseases. Since the 1940s, Tanzania had also begun to export small quantities of dried seaweeds of the genus *Eucheuma*. But seaweed farming in Tanzania – or, for that matter, in Africa – did not exist.

In my studies, I also learned that seaweeds, in addition to their edibility and nutritional and medicinal value, could be processed into a wide variety of products, including salad dressing, animal feedstock, fertilizer, toothpaste and shampoo. Most importantly, I observed how seaweed often continued to grow and thrive when fragments (fronds) were torn from their original habitats to become wedged into rock outcroppings just beyond the shoreline. For me, this meant that at least some species of seaweed could likely be cultivated using vegetative fragments as planting material.

Today, Tanzania exports nearly 15,000 tonnes of *Eucheuma* seaweed each year, largely from the islands of Zanzibar, Pemba and Mafia. This compares to just 500 tonnes exported annually in the late 1960s and early 1970s when the tonnage was derived solely from wild seaweed biomass harvests.

I published my first scientific paper on Tanzanian seaweeds, "The Economic Importance of Seaweeds: Can These Plants Contribute to the Economy of Tanzania?" in the *Journal of the Geographical Association of Tanzania* in July 1969. I published a second paper, "Exploitation of seaweeds in Tanzania: the genus *Eucheuma* and other algae", in *Tanzania Notes and Records* in 1973. I published a third paper, "Seaweed farming in Tanzania: a possibility in Tanzania's coastal *ujamaa* villages", also in *Tanzania Notes and Record,* in 1976.

Strangers, then friends

Salam was likely to have learned about these publications and also about my subsequent writings while visiting Tanzania in 1984.

During this period, my research on seaweeds received funding from the International Foundation for Science (IFS) of Sweden and the US Agency for International Development (USAID). It was also during this period that Thomas Odhiambo, a renowned entomologist at the University of Nairobi in Kenya, founded the International Centre of Insect Physiology and Ecology (ICIPE).

Odhiambo was deeply interested in harnessing science – particularly agricultural science – to improve economic and social conditions across Africa. Like Salam, he was an eminent scientist and a dedicated humanitarian and reformer. He devoted his life's work to the proposition that science and social well-being were one and the same. He was a founding member of TWAS, and also a member of the TWAS Council. Though I did not know him until I became a TWAS Fellow, Odhiambo is also likely to have been one of the scientists responsible for my election to TWAS in 1987, and for my election to the African Academy of Sciences (AAS) a year later.

Like Salam, when Odhiambo heard about a scientist doing important and influential work, he was quick to render his support to that person. His efforts on my behalf again illustrate how the kindness of strangers (who often later became friends) has played such an important role in my career.

My research tracked closely with Odhiambo's quest for devising deeply rooted, scientifically verifiable, agricultural practices to help Africa's economy grow for the benefit of all the people. As much as anyone, Odhiambo opened the doors of TWAS to me and helped me see the Academy as an agent of social change. In this context, I recall travelling with him from Nairobi to Butiama village in early 1994 to visit President Nyerere, who had recently stepped down as president of Tanzania and was living in the village of his birth. There we discussed strategies for securing sustainable funding for research and development in Africa.

Milestone years

The year 1969 served as an important milestone in my career for a number of reasons. After being accepted to several universities for graduate studies, I chose to attend the University of Hawaii in the United States, where I would earn a doctorate degree in botanical sciences in 1974.

I went to the University of Hawaii because Maxwell Doty, one of the world's most eminent seaweed researchers, was there. His work focused on the tropical seaweed of the genus *Eucheuma*.

Doty, who was about to launch a number of seaweed farming trials in the Philippines, helped me to secure a travel grant from Marine Colloids Company, a US-based corporation that was interested in increasing the commercial value of *Eucheuma*. The grant financed a two-month trip to the Philippines, where together with Doty's other graduate students, I conducted field experiments and farm trials as a prelude to my studies in Hawaii.

I will be forever grateful for the opportunity to study abroad, which was made possible by a fellowship from the Rockefeller Foundation. The doctoral programme at the University of Hawaii allowed me to broaden both my base of botanical knowledge and research skills. It also opened my world to new experiences and insights that have enriched both my life and career.

Nevertheless, I had no intention to remain in the United States upon earning my degree. In fact, I left Honolulu for Dar es Salaam just 48 hours after defending my dissertation. I was confident, inspired, excited and, most of all, eager to share my knowledge on seaweeds with the people of Tanzania. I was the first PhD graduate in marine botany in my country.

Upon returning home, I was appointed a lecturer in botany at the University of Dar es Salaam. I swiftly climbed the academic ladder, eventually becoming a full professor and founding director of postgraduate studies at the university.

I would subsequently serve as founding pro-vice-chancellor for academic affairs and research at the University of Namibia, and, more recently, as vice-chancellor of Hubert Kairuki Memorial University, Tanzania's first private university initially devoted to training medical doctors and nurses. I was also founding chair for promoting the United Nations Educational, Scientific and Cultural Organization-United Nations University (UNESCO-UNU) Zero Emissions Research Initiative (ZERI) in Africa; chair of UNU's Advisory Council on ZERI; and a member of a UN Expert Group on Climate Change and Sustainable Development.

The benefits of membership

I was elected to TWAS in 1987 and quickly became an active member of the Academy. Since then, I have attended virtually all of the Academy's annual meetings. I also chaired the agricultural sciences election committee from 1993 to 1997.

I owe the Academy a great deal. TWAS was instrumental in helping me win the African Academy of Sciences (AAS)-Chemical Industries Basel (CIBA) prize for Agricultural Biosciences in 1993 and it supported my quest to earn the Boutros Boutros-Ghali Prize in 1994, in recognition of my contributions to scientific research and development in Africa.

Mohamed Hassan, TWAS's founding executive director, was a particularly energizing force on my behalf. He not only urged me to compete for these prizes but also maintained confidence in me throughout the selection process. Winning these prizes helped to raise my profile in the international scientific community. Such notoriety, in turn, opened up additional opportunities in subsequent years.

Similarly, my long tenure (1994 to 2010) as editor-in-chief of *Discovery and Innovation*, the flagship journal of the African Academy of Sciences (AAS), came to be, in part, because of my close ties with TWAS and AAS. For these opportunities, I extend my sincere thanks to Odhiambo, who became one of my primary mentors.

These are just a few examples of how the contacts I established with colleagues across the developing world and especially in sub-Saharan Africa – as a result of my involvement TWAS – helped to broaden the scope and impact of both my research and outreach efforts.

Over the years, TWAS has helped to enrich my career in untold ways through friendships and partnerships with many scientists that I have come to know. Indirectly, the Academy has also helped to rekindle my interest in mushrooms and expand my involvement in science education. I have truly cherished my association with TWAS.

Pursuing Salam's vision

The Academy has experienced unprecedented growth in the past decade. The number of TWAS members has increased by nearly 30%. TWAS's fellowship programme, currently offering more than 500 grants each year, has become one of the largest in the world. At the same time, funding from various organizations – most notably, the Swedish International Development Cooperation Agency (Sida) – has more than doubled.

The Academy has also emerged as one of the world's leading advocates for scientific capacity building and science-based development in the developing world. This is no small feat given the complexity of the task and the large number of institutions with similar mandates and goals.

TWAS may have been a lonely voice for building scientific capacity and excellence in the South when it began 30 years ago, but that is no longer the case, either at a national or international level. TWAS deserves to be congratulated for standing out in a crowded field and, even more importantly, for the unique contributions it has made to science and development.

Despite this progress, key challenges must be met if TWAS is to fully realize the vision that Abdus Salam first presented when he launched the Academy three decades ago. I would like to focus on three major challenges that the Academy should seek to address in the years ahead.

First, there is the 'ageing' challenge. Science academies, including TWAS, recognize the accomplishments of the world's most eminent scientists, often honouring individuals who have acquired their well-earned standing over the course of long and distinguished careers. Most science academies, as a result, consist of older scientists who have been chosen because of achievements attained over decades. TWAS is no exception.

It should be noted that the Academy has done an excellent job of expanding its fellowship programme for young scientists. It has also taken significant steps to introduce the Academy to the next generation of scientists in the developing world through its Young Affiliates programme, which has enabled promising scientists under the age of 40 to participate in TWAS activities.

I became a member of TWAS at the relatively young age of 43. My selection, however, seems to be the exception, not the rule. The average age of TWAS Fellows currently is 70.

TWAS should therefore continue pursuing an active programme to attract and engage younger scientists. An inevitable generational change is under way, and the Academy should welcome and encourage this trend. It should even consider setting aside a few openings for the election of worthy young scientists each year. Merit, of course, must remain the most important criteria. But opening up the Academy to young scientists will likely benefit both TWAS and, more generally, science in the developing world. I am an example of how valuable membership at a young age can be. Mohamed Hassan's example is even more convincing. He was elected a TWAS member at the age of 38.

Second, TWAS should continue taking pragmatic steps toward increasing the number of women scientists among its members. Progress has been made in achieving greater gender equity within the Academy. A decade ago, women comprised less than 3% in the membership. Today, women comprise 10% of the membership. That is not good enough, but I am optimistic.

Part of my optimism is due to the rising number of women in developing countries who are pursuing careers in science, especially in biology, ecology and medical research (less so in physics, engineering and mathematics). Yet part of this increase is also due to the Academy's dedicated efforts to identify female scientists who meet the qualification criteria for membership. While TWAS should be commended for vigorously addressing the issue of gender imbalance, it must not become complacent. Over time, women should represent half of TWAS's membership.

Throughout the Academy's history, just five women have served on the TWAS Council: Lydia Makhubu, Eugenia Maria Del Pino Veintimilla, the late Farida Shah, Fayzah Al-Kharafi and Rabia Hussain. While notable progress has been made in recent years in these appointments, it is my fond hope that multiple female scientists on the Council will become the norm in the years ahead. As with its membership, TWAS should only claim its efforts to achieve gender equality a success when half of the Council is comprised of women.

Third, TWAS must continue seeking ways to address the trends that have tipped its membership and awards towards scientists in a select group of countries. While the Academy has achieved unparallelled success in honouring and rewarding scientific excellence in the developing world, its membership of nearly 1,150 Fellows is heavily skewed toward the largest, more advanced developing countries – notably Brazil, China and India. The same three countries tend to dominate TWAS awards and prizes.

Only four of the 52 scientists elected to TWAS in 2013 hailed from Africa (Benin, Ethiopia, Kenya and Tanzania). African scientists thus comprised only 8% of the class of 2013. Conversely, nine new members came from China, and another four from Taiwan, China; 11 came from India; and another 11 from Brazil. If we add the new members elected from Australia, France, Japan, South Korea, the United Kingdom and the United States, then inductees to the Academy's class of 2013 from emerging economies and developed countries accounted for 80% of the total. The same trends are evident in the selection of TWAS prizewinners: 12 of the 16 recipients were from Brazil; China; Taiwan, China; and India.

The TWAS Council is very aware of this challenge and has rigorously discussed how it might be addressed. The difficulty is rooted in the Academy's twin – and sometimes conflicting – goals to build scientific capacity throughout the developing world and to honour scientific excellence among individuals who have attained the highest levels of achievement.

The duality between equity and excellence has been at the heart of TWAS's challenge since the Academy's launch. It was reflected in the large percentage of the Academy's founding members who were working in institutions in the North at the time of TWAS's inception, and by Salam's acknowledgment that the standards of excellence for science had been set by the developed world (and should be adopted by the South). When Salam spoke about the universality of science, he was primarily referring to a universality that had been most forcefully expressed by countries of the North over the preceding 500 years.

It is one thing to passionately discuss the impact of a North-South divide in science, and quite another to directly confront the ramifications of a South-South divide in science. Forces well beyond the reach of TWAS, of course, are driving these trends. Yet, it is also true that forces well beyond the Academy's reach have largely driven the North-South divide in science.

What obligation does TWAS have to narrow the South-South gap in science, and what strategies could best be employed to address this growing critical concern?

The Academy's programme to support research groups in the least science-proficient countries is a prime example of some of the steps being taken. The dramatic increase in the number of fellowships for PhD and postdoctoral study, which include a sizeable number of young scientists from Africa, are welcomed and appreciated. So, too, are the C.N.R. Rao Prize in Scientific Research and the Atta-ur-Rahman Prize in Chemistry for researchers from scientifically lagging countries.

But these measures remain too small in scope to address this fundamental issue: How can TWAS quicken the pace of capacity building in scientifically lagging countries, primarily in sub-Saharan Africa and countries with predominantly Muslim populations? Stated even more simply: How can the issues of equity and excellence become one and the same throughout the developing world? The answer lies in raising the playing field in countries that have lagged behind in scientific capacity.

This challenge might be addressed by investing additional funds for improving scientific infrastructure in scientifically lagging countries; by providing an even larger number of PhD fellowships to young African scientists; by extending greater support to African scientists during the early stages of their careers; and by recognizing the difficulties that African scientists face, perhaps by granting bonus points in the election of members and the awarding of grants and prizes.

Some may call the latter proposal "affirmative action" and worry that it poses a risk to the Academy's strict focus on excellence.

Conversely, it might be considered a nod toward flexibility that could help TWAS address one of its more vexing challenges.

The Academy, of course, needs to continue to emphasize scientific excellence. Yet it must also acknowledge that, while scientific talent resides everywhere, conditions dictate that excellence is easier to achieve in some places than in others.

Sustaining success

The South-South gap in science is likely to continue to be a major challenge confronting TWAS, as it moves beyond the success it has achieved over the past 30 years and as science in the South continues to advance at an unprecedented but uneven pace.

In some ways, this challenge is a reflection of the progress that has been made – at least in some countries. As a result, measures of success for TWAS are no longer likely to be solely determined by closing the gap in science between the North and South. It is becoming increasingly important to narrow the gap within the South as well.

No other organization is more qualified to guide this effort than TWAS. That is why I am convinced that the Academy will remain as relevant in the future as it has been in the past.

The focus and strategies of TWAS may change over the next 30 years. But its goals are likely to remain true to the original concepts first articulated by Abdus Salam at the Academy's inception: to build scientific capacity and to recognize and reward scientific excellence, throughout the developing world. From my view, now that a portion of the journey has been achieved, it makes the remainder of the quest even more exciting and attainable.

A life of change

Yu Lu

Yu Lu, a world-renowned Chinese physicist, was born in 1937 in Zhenjiang, a city on the Yangtze River in southeast China, where he was raised and received his early education.

Based on his academic achievements, he was selected to attend Kharkov State University in the Soviet Union, graduating in 1961. Upon returning to China, he was appointed a researcher at the Institute of Physics, which is part of the Chinese Academy Sciences (CAS).

Yu Lu planned to return to Kharkov University in the early 1960s to earn a doctorate degree in physics. However, political tensions between China and the Soviet Union prevented him from doing so. Disappointed but not despondent, he continued his research at the Institute until the Cultural Revolution led to his forced exile in rural China. There he toiled as a manual labourer for two years from 1969 to 1971, before returning to CAS to resume his research career.

In 1978, as China began opening to the rest of the world, he was given an opportunity to travel to Europe. One year later, he became a fellow at Harvard University, where he remained for 16 months, followed by a six-month stay at the University of California, Santa Barbara (UCSB).

In 1986, Yu Lu was appointed the first permanent member of the International Centre for Theoretical Physics (ICTP) scientific staff, and later, in 1990, the head of Centre's condensed matter group. He would remain at ICTP until 2002, serving as the driving force behind the development of the condensed matter group.

In 1990, Yu Lu was elected a fellow of TWAS. Over the next decade, he played a critical role in helping the two organizations develop collabora-

tive activities designed to advance their shared goals for building scientific capacity and excellence in the developing world.

Upon leaving ICTP, Yu Lu returned to CAS, where he has pursued a rigorous research agenda as one of the nation's preeminent physicists. He has also been active in China's efforts to forge strong international links with scientific communities in both the South and North.

Yu Lu has published seminal articles in a broad range of research fields, including strongly correlated systems, superconductivity and quantum statistical field theory. He has received multiple awards from CAS and the Chinese government, including the CAS Prize in Progress of Science and Technology, the CAS Prize in Natural Sciences, the National Prize in Natural Sciences and the National Prize in Progress of Science and Technology. In addition, he was a recipient of the Institute for Scientific Information (ISI) Classical Citation Award in 2000 and the American Institute of Physics' John T. Tate Medal for International Leadership in Physics in 2007.

In the following essay, Yu Lu tells a fascinating personal story. His journey both tracks and intersects with the development of ICTP and TWAS over the past three decades. It speaks to the enormous changes that have taken place in science not only in China, but also in a growing number of other countries – changes that are affecting the lives of tens of thousands of scientists in the developing world.

***

I was born in 1937 in Zhenjiang, a city located on the Yangtze River in southeast China, into a closely knit, happy family of modest means.

The Nationalist government of Chiang Kai-shek was in power during much of my early childhood, but a great deal of my education – most notably, my secondary school education – took place after the rise of the People's Republic of China in 1949.

I had the good fortune of being an excellent student, especially in the sciences. I was even more fortunate to have my talents recognized. Based on my scholastic records, I was selected by the government to attend Kharkov State University in the Soviet Union.

The 1950s were a time when many of China's most promising students went to the Soviet Union for their university and postgraduate education.

Except for a few scientists trained in the West in the 1940s or earlier, Soviet-trained students like myself constituted a significant portion of the first generation of researchers in post-revolution China.

Today, as we approach and enter retirement, our numbers are dwindling. Nevertheless we still represent an important segment of China's research community.

After graduating from secondary school in 1954, I spent two years learning Russian in preparation for my studies in the Soviet Union. Kharkov was the third most important centre for science in the Soviet Union, exceeded in reputation only by universities and research institutes in Moscow and St. Petersburg (which was then called Leningrad).

I received an excellent education there and was able to interact with eminent faculty members and outstanding fellow students. It was a challenging and rewarding learning environment that provided a firm intellectual foundation for my subsequent research. The experience proved instrumental in shaping my career.

The perils of isolation

Upon earning an undergraduate diploma in physics in 1961, I returned to China to take a position in the Institute of Physics at the Chinese Academy of Sciences (CAS). Established in 1949, immediately after the creation of the People's Republic of China, CAS is the country's largest and most prestigious scientific institution.

At the time, China's research facilities were poorly equipped and researchers were largely isolated from their counterparts, especially in the West. This was true in virtually all ways except one: We had reasonably good access to current literature. Under difficult circumstances, our libraries were able to acquire up-to-date journals and books. As a result, we could keep current with some of the latest developments in our research fields.

Such access enabled us to conduct interesting research and to remain productive in our chosen disciplines. This was particularly true for mathematicians and theoretical physicists who worked in fields that did not require expensive equipment. In this sense, China confirmed Abdus Salam's notion that mathematics and theoretical physics were ideal research areas to promote in developing countries because they were cheap. From an historical perspective, China's experience verified Salam's contention.

I do not want to minimize the challenges we faced. In the early 1960s, China was an extremely poor country, investments in science were minuscule, scientists were few in number, and the integration of scientific research into the larger society remained tenuous at best.

Although we had access to scientific literature, a lack of personal exchange with other scientists meant that we could not possibly know the range and depth of research taking place elsewhere. Even more importantly, we simply did not have the opportunity to acquire and test our knowledge through direct interactions with foreign colleagues. In effect, we lacked one of the primary channels for making scientific progress.

At the same time, it is also true that scientists in other countries, even communist countries with close ties to the Soviet Union, remained unaware of the nascent research that was taking place in China. Ignorance on both sides of the East-West divide generated a mutual sense of surprise when the Chinese scientific community opened its doors to the global scientific community at the second TWAS General Meeting in Beijing in 1987.

As a young researcher, I was satisfied with both the diversity and depth of my research activities at CAS in the early 1960s. Within the limitations of our research environment, we were able to do stimulating and challenging work. For example, soon after my return from Kharkov, I was encouraged to pursue cutting-edge research in the emerging field of superconductivity.

I was just 24 years old and did not have an advanced degree. My diploma supervisor strongly recommended that I return to Kharkov State University to earn a doctorate in physics. I was very excited about the challenging work ahead and the prospect of advancing my education and career.

Rift and retreat

Unfortunately, my ambitions were abruptly derailed by growing tensions in Chinese-Soviet relations, which halted all bilateral scientific exchange programmes. The political rift first surfaced with the withdrawal of Soviet scientists, engineers and technicians from China in the early 1960s.

Their withdrawal was a manifestation of deep divisions in ideology and competing national interests, problems that had been festering for some time. Geopolitical strains between the Soviet Union and China had translated into a dramatic alteration in my career trajectory.

It was a confusing time for me. As a reward for being an excellent student, the government had sponsored my education at Kharkov State University, presumably to become a more productive citizen when I returned home from my studies.

Now, political disagreements between China and the Soviet Union not only short-circuited my return to Kharkov to complete my education but also raised suspicions about my loyalties – at least to the extent that I was denied travel not just to the Soviet Union but to all foreign countries. Little did I realize at the time that the travel ban would last 17 years.

Accepting my fate, I continued to work at CAS as a researcher, where I was able to conduct my studies at a satisfactory level. In 1965, I published a path-breaking paper on superconductivity in a Chinese scientific journal. The article would remain unknown in the West until it was brought to the attention of foreign colleagues 30 years later.

This incident conveyed two salient points: First, cutting-edge research in theoretical physics was possible in China, at least to some extent, even at this early date and, second, the findings went unnoticed because of China's isolation.

Then, another tumultuous national event intervened to directly impact my career.

Like many other intellectuals and scientists in China, I was banished to the countryside to be "re-educated" during the Cultural Revolution, which lasted from 1966 to 1976. My exile to a remote rural area of the country, with virtually no access to the outside world, began in 1969 and ended in 1971.

During this period, I was entirely cut off from the scientific world, doing manual labour and living in quarters that had previously been part of a labour camp for prisoners. At the conclusion of this dark period, I came back to Beijing to re-join CAS.

Luckily, the good fortune that had accompanied the earliest years of my career surfaced once more. I had access to current scientific literature and started, once again, to pursue a fairly sophisticated research agenda. I quickly re-engaged with the Chinese science community and resumed my research in condensed matter physics and superconductivity.

The research that I had carried out during this period would eventually provide a passageway to the international scientific community and, more specifically, to ICTP in the mid-1980s.

Most people are products of their times, yet their lives are largely divorced from the seminal events of history. People are often only marginally impacted by the sweeping, society-altering events of their era.

For better or worse, that has not been the case in my life. Sino-Soviet agreements for scientific exchanges, founded on their shared principles for governance, led to a government-sponsored exchange programme that provided me with an excellent university education and a strong intellectual foundation for my career as a theoretical physicist. Then, the fallout between China and the Soviet Union, based on different perceptions on how to enact communist principles, abruptly ended my quest for a doctorate just as the arrangements for my studies had been finalized. Later, the Cultural Revolution, launched by Mao Zedong to purify China of corrupting influences, interrupted my work as a researcher and led to an imposed migration to the countryside.

The flow of history

I have sometimes been in the right place. I have sometimes been in the wrong place. Yet the irony is that I was usually in the same place, serving as both an observer and recipient of the dramatic events that cascaded across China over the last half of the 20th century. For me, seminal historical events have had direct consequences.

This pattern continued in the late 1970s and early 1980s when China's efforts to open up to the rest of the world again led to fundamental alterations in my life and career.

The economic opening of China began in 1978 under the leadership of Deng Xiaoping. Free trade zones were established, foreign investment was welcomed and entrepreneurship was encouraged. Scientific research was again viewed as a pursuit of high value to society. China, as a result, was transformed in a matter of decades.

The initial efforts to strengthen China's scientific capacity, which can be traced to the early 1970s, culminated in the TWAS's second General Meeting held in Beijing in 1987, when China's scientific community opened up to the world for the first time.

The events leading to the TWAS conference began in 1972, when United States President Richard Nixon visited China, ending a quarter-century of icy separation between the two nations. That same year, TWAS Associate Founding Fellow C.N. Yang, a world-renowned Chinese-American physicist and Nobel laureate in physics, also visited China at the invitation of Chinese officials.

Throughout this trip, Yang emphasized the importance of basic science in promoting China's long-term economic growth. He offered his opinions solely as a scientist. This insulated his remarks from the political arena and thus made them less controversial. His focus was on the positive impact that science has had – and will continue to have – on development in all countries. His views were well received by the Chinese leadership.

Three years later, in 1975, a delegation from the American Physical Society came to China, again to discuss the importance of science to society in a non-political context. Four Nobel laureates, including J. Robert Schrieffer, who would later play a critical role in forging my connections to ICTP and TWAS, were among the participants. Their message was also well received.

Upon returning to the United States, the visitors produced a 'blue book' on the state of physics in China that was made available to the international scientific community. It offered other countries one of their first glimpses of the state of science in my country.

These two high-profile events gave a significant boost to science in China, marking the beginning of a new era of support for research that has continued non-stop ever since.

Science openings

History shows that a scientific opening actually preceded China's economic opening. In the 1980s, the two reform movements joined together to play a seminal role in the dramatic development of China that followed.

Over the next three decades, more than 600 million people in China have been lifted out of poverty and the country's economy has expanded, on average, by nearly 10% per year.

In 1978, in the aftermath of the unprecedented scientific exchanges that took place earlier in the same decade, I was offered an opportunity to travel to Europe. It was my first trip outside of China in nearly two decades. The following year, in 1979, I became a visiting scholar at Harvard University, where I worked with Bert Halperin, who was four years younger than me. He often jokingly referred to me as a truly senior fellow.

Here, I began the second significant phase of my education in physics. The first had taken place in the Soviet Union in the late 1950s and early 1960s when I was in my 20s. The second phase took place in the United States when I was in my 40s. I would spend 16 months at Harvard and an additional six months at the University of Santa Barbara Institute of Theoretical Physics in California. Unlike my experience during the Cultural Revolution, this was a true re-education.

During the next few years, the pace of my travel picked up considerably as I attended conferences and workshops in the United States and Europe. During these years, I got to know Stig Lundqvist, largely by way of an introduction from Schrieffer.

Lundqvist was not only a good friend of Schrieffer's, but he also worked closely with Abdus Salam, the founding director of ICTP in Trieste. His involvement in ICTP dated back to the late 1960s and, from 1984 to 1992, he chaired the ICTP Scientific Council. Lundqvist was the driving force behind the creation of the Centre's condensed matter group, one of the ICTP's first ventures beyond the field of theoretical high-energy physics.

Lundqvist invited me to the Centre in 1983. There I spoke to Abdus Salam in person for the first time. I had seen him from a distance 20 years earlier, in 1964, during his first visit to China, when he attended a science forum. I was one of many scientists in the audience when Salam gave a presentation at CAS. By then, Salam was already a revered figure in the South. I remember how inspiring it was to be in the presence of the world-renowned scientist from Pakistan, a country that was as poor as China at the time.

Short visits, lasting stay

In 1983, the same year that TWAS was founded, I became an ICTP Associate member, which allowed me to visit the Centre periodically for relatively extended periods. In 1985, Salam sent me a letter asking if I would consider moving to Trieste to become head of ICTP's Condensed Matter Group. The letter was both a surprise and an honour. I suspect that Schrieffer suggested to Lundqvist that I would be a good candidate to head the group.

I was ICTP's first permanent researcher. Before then, the Centre had filled positions with faculty from universities and research institutions in Italy and elsewhere. The faculty would devote a percentage of their time to organizing and teaching the Centre's seminars, workshops and schools.

After arranging a two-year leave of absence from CAS, I accepted the offer. I arrived at ICTP in 1986. Like so many others who came to the Centre, my short-term appointment would turn into a long-term commitment. I remained at ICTP for 16 years, until 2002.

The focus throughout my stay in Trieste was ICTP. Nevertheless, during the early years, I closely interacted with TWAS, which had just opened its doors on the campus of the ICTP. The relationship would mature and expand over time, especially as I got to know Mohamed H.A. Hassan, the Academy's executive director.

In 1990, I was elected a TWAS Fellow, one of the most important honours that I have received in my career. It is one that I particularly cherish because it was bestowed by my peers from the across the globe.

A positive partnership

I am not the first to note that ICTP played a critical role in TWAS's development and success. It is well known that Abdus Salam was largely responsible for the creation of both organizations. In fact, he envisioned ICTP and TWAS as complementary organizations – the former dedicated largely to training researchers from the developing world and the latter focused on honouring scientists who had made significant contributions to their research fields.

ICTP has served as a critical resource for building the Academy's membership, especially during TWAS's earliest years. Some of TWAS's most distinguished members, most notably those from Africa and small developing countries, were actively involved in ICTP – for example, Francis Allotey from Ghana and Virulh Sa-Yakanit from Thailand. These scientists, upon their election to TWAS, became associated with one of the world's preeminent international science academies, at once reinforcing and expanding the strengths of both ICTP and TWAS. From the start, it was a relationship that generated mutual benefits.

The two organizations developed complementary strategies to advance their shared goals. ICTP, for example, established a Training and Research in Italian Laboratories (TRIL) programme that enabled scientists from developing countries to visit Italian research institutions both for training and engaging in joint projects. TWAS adopted a similar initiative, but placed it on a South-South template, allowing scientists from less scientifically proficient countries to visit centres of excellence in more scientifically proficient countries in the developing world, both for training and collaboration.

In the same vein, ICTP's research capacity building efforts were focused on mathematics and physics, while TWAS's capacity building efforts spanned the full spectrum of scientific disciplines, although the Academy has often deferred to ICTP when addressing challenges related to mathematics and physics.

ICTP has generously provided office space and administrative assistance to TWAS to help defray the Academy's overhead expenditures. Extending this helping hand proved especially beneficial during the Academy's early years when its financial resources were extremely tight. It continues to be a source of cost-savings to this day.

Side by side

I would be remiss if I didn't note that ICTP has benefited from TWAS's presence on its premises. Side-by-side secretariats have strengthened the organizations' mutual goals. The close relationship has served both organizations well in good and bad times.

For example, during the financial crisis that struck ICTP and TWAS in the early 1990s, it was not clear whether the Italian government would continue to provide funding. The combined activities of ICTP and TWAS helped to highlight how significant the loss would be – not only to international science, but to Italy as well – in terms of the good will and the deep sense of kinship to Italy that these programmes had engendered across the developing world. It is not simply rhetoric when scientists call Trieste their "home away from home". Rather, it is an expression of gratitude for the help that the international scientific institutions in the city have given them throughout their careers.

In China (and elsewhere), researchers and public officials often do not distinguish between ICTP and TWAS. Rather, they think of the organizations as integrated components of the Trieste science system dedicated to the support of scientists in the South.

ICTP and TWAS have engaged in an increasing number of collaborative programmes over the past couple of years, including jointly funded workshops and conferences. This has not only reinforced their long-standing ties but it has also helped to advance their common goals. As previously mentioned, Salam viewed ICTP and TWAS as complementary institutions. Their proximity to one another and close working relationship has made each organization much stronger than they could have ever been on their own.

Science synergies

The state of global science has changed dramatically over the past three decades. The pace of change, moreover, has been accelerating.

Many factors account for the unprecedented growth in scientific capacity and excellence that has taken place in the South: National investments in science have increased enormously among a growing number of emerging economies and developing countries; rising wealth among countries in the South (based in part on their expanding scientific capabilities) has led to even greater investments in science and technology; South-South collaboration in science (one of Salam's highest priorities) has grown, enabling countries to learn from one another; and the North-South gap in scientific capacity and excellence has narrowed, improving the prospects for research initiatives that are international in scope and impact.

It is interesting to note that the history of ICTP, which celebrated its 50th anniversary in 2014, and the history of TWAS, which celebrated its 30th anniversary in 2013, parallel the enormous changes in global science that have taken place during this period.

ICTP's annual budget, which is currently USD20 million, and TWAS's annual budget, which stands at USD5 million, are too small to change the world. Yet, the role of both organizations – both individually and collectively – in helping to spark and then fuel the dramatic changes in science in the South should not be underestimated.

Many of the strategies that have been used to build scientific capacity and excellence in the South were first introduced at ICTP and TWAS, including associateship programmes, short-term workshops and conferences designed to overcome the isolation that scientists in the developing world have experienced; PhD and postdoctorate fellowship programmes for ensuring that promising young scientists could continue their careers; and a broad array of forums for exchanging ideas and networking that have helped science in the South to become as vital and innovative in the developing world as it has been in the developed world.

These programmes have not only assisted thousands of scientists in their own right, but have also served as prototypes for building scientific capacity through bi-lateral and multi-lateral arrangements among individual nations and international organizations.

ICTP and TWAS have been key players in changing the face of global science. In the process, they have helped to produce sustainable economic development in a growing number of emerging economies and developing countries.

New dynamics

No country has come to symbolize these changes more than my own country. Today, China is the world's second largest economy. It has developed world-class research facilities in a growing number of fields, including particle physics, condensed matter physics and materials science.

Articles published by Chinese scientists in peer-reviewed scientific journals now account for more than 10% of the articles published worldwide, and the number is rising at a rate of more than 20% each year. In some fields – for example, nanotechnology – Chinese scientists rank No.1 in publications.

While China's increasing scientific output may be more impressive than that of any other developing country (this is a case where the breadth of the policies and the size of the population have together made a difference), trends in scientific publications in China are not an anomaly. The number of articles published in peer-reviewed scientific journals by researchers in a growing number of emerging economies and developing countries – for example, Brazil, India, Iran, Malaysia, Mexico and Turkey – is also increasing at a rapid pace.

Such positive developments provide ample reason to celebrate the progress that has been made. Yet it also calls for a fresh look at ICTP's and TWAS's efforts to ensure that the institutions remain as relevant in the future as they have been in the past. ICTP and TWAS have helped the world move on, and both institutions must continue to move on in order to keep pace with the changes that they have helped to create.

I think there are three key measures that ICTP and TWAS can take – both individually and together – to enhance their capacity-building efforts in the future.

First, more scientifically advanced countries in the South should provide greater support for developing countries that are lagging behind. I think that both ICTP and TWAS have taken steps to address this issue by nurturing closer ties with scientifically strong countries in the South. The election of Bai Chunli, president of CAS, to the presidency of TWAS, promises to accelerate this trend.

Second, South-North cooperation should shift its emphasis from a paradigm based on a superior, more capable North lending a helping hand to an inferior, less capable South to one that emphasizes shared capabilities, shared responsibilities and shared benefits. The North obviously had more to offer than the South in collaborative projects in the past. But that may no longer be the case. The South now has valuable insights to convey to the North not just in terms of its research, but also in terms of national investment strategies for building scientific capacity. In some cases, such strategies appear to be more effective than the strategies currently being pursued by developed countries.

Third, scientific communities should not only focus on building scientific capacity but also on ensuring scientific excellence. A dedication to excellence was at the heart of Salam's strategy for improving science in the developing world at the time of ICTP's and TWAS's creation. This goal remains as important today as it was then. Excellence lends credibility to a nation's research agenda, it helps sustain funding and it is essential for international collaboration. Scientific capacity building without excellence is unsustainable.

A great deal of progress has been made on this front. First-class research is now taking place in a growing number of fields in an increasing number of countries in the South.

Completing the journey

Yet, the journey is by no means complete and now is no time for proponents of science for development to rest on their laurels. The focus must remain on both excellence and fair play, which includes respect for intellectual property and proprietary information. Part of the quest for excellence includes abiding by the rules – an essential factor if international collaboration is to continue expanding.

Thirty years from now, on the occasions of TWAS's 60th anniversary and ICTP's 80th anniversary, I hope that the recent paradigm-altering trends in global science will have come to complete fruition. The North, I strongly believe, will be producing its fair share of global science but it will not hold its once dominant position. The South, meanwhile, will no longer play a subsidiary role but will be contributing to international science as a full and equal partner.

As a result, a large number of countries will be making fundamental contributions to basic science and playing significant roles in addressing the world's most pressing and complex environmental and health issues. Poverty will be afflicting an ever-smaller percentage of the world's population, thanks in large measure to advances and applications of science and technology. The focus of the global scientific community will have shifted to addressing our most intractable issues: climate change; water, food and energy security; biodiversity loss; and the risks posed by natural disasters and the spread of infectious diseases. Formidable challenges will undoubtedly still be with us – perhaps even more formidable than today. But science, now being done by more people in more places, will be better positioned to deal with them.

From this perspective, I am optimistic. It is optimism drawn not only by casting my eyes towards the future but also by peering back at my own career – a career that shows that anything is possible, many times over, especially with the support of forward-thinking and highly competent institutions like ICTP and TWAS.

Moving ahead together

Roseanne Diab

Roseanne Diab is the executive officer of the Academy of Science of South Africa (ASSAf) and professor emeritus in environmental sciences at the University of KwaZulu-Natal.

Diab is internationally recognized for her research in the atmospheric sciences, especially for her studies of air pollution, air quality management and climate change. She is also well known for her advocacy for science in the developing world, especially in South Africa, and for her promotion of gender equality in science.

She has more than 100 academic publications to her credit and has served on the editorial boards of the South African Journal of Science, Clean Air Journal, South African Geographic Journal *and* Atmospheric Environment. *Her research has often focused on atmospheric conditions and trends in Africa.*

Born and raised in Durban, South Africa, Diab received her primary and secondary education in her hometown. She earned a bachelor of science degree in 1969 and Master of Science degree from the University of Natal (now the University of KwaZulu-Natal) in 1975. She then journeyed to the United States, where she obtained a doctorate degree in environmental sciences from the University of Virginia in 1983. Following a distinguished career as a professor of environmental sciences at the University of KwaZulu-Natal, she was named the executive officer of ASSAf in 2008.

During her tenure as executive officer, she has sought to raise the public profile of ASSAf and to highlight the important role that the academy plays in examining critical science-related issues in South Africa, especially those pertaining to education, sustainability and health.

Diab has been a Fulbright Senior Fellow and a recipient of a Rockefeller Foundation Bellagio residency award. She is a member of ASSAf and a fellow of the South African Geographical Society and the University of KwaZulu-Natal. She has served on a number of international bodies, including the International Ozone Commission and the Commission on Atmospheric Chemistry and Global Pollution. She was a member of SPARC (the Scientific Steering Group of the Stratospheric-Tropospheric Processes and their Role in Climate) in the World Climate Research Programme. She has also been active in the Organization for Women in Science for the Developing World (OWSD).

In 2010, Diab co-chaired the InterAcademy Council's (IAC) study panel that was responsible for reviewing the procedures and administrative structure of the Intergovernmental Panel on Climate Change (IPCC). Its influential report, Climate Change Assessments: Review of the Processes and Procedures of the IPCC, *received international press coverage and became a main reference source in discussions within IPCC.*

Diab was elected a TWAS Fellow in 2011. She currently serves on the Academy's astronomy, space and earth sciences membership advisory committee and is a member of the Academy's gender advisory board.

In the following essay, Diab explains how she chose to pursue a career in science and to develop a research specialty in the atmospheric sciences. She then explores the major challenges facing South Africa's scientific community today. She concludes by describing the nature of her involvement in TWAS as she has become increasingly engaged in issues related to gender and youth that promise to be central to the Academy's agenda in the years ahead.

***

Before being named executive officer of the Academy of Science of South Africa (ASSAf) in 2008, I was unaware of TWAS. But that would change rapidly over the next few years.

My initial knowledge of TWAS was confined to its role as a merit-based science academy designed to recognize the most eminent scientists in the developing world. During my first months at ASSAf, when someone mentioned TWAS, what immediately came to mind were its members. That is the aspect of the Academy that fostered my familiarity with TWAS during my early tenure with ASSAf.

It would take me a bit longer to learn about TWAS's broad range of programmatic activities – its fellowships, awards and prizes – as well as the prominent role that the Academy has played as a voice for science in the developing world.

First impressions

In a sense, my very first impressions of TWAS did not do full justice to the Academy's expansive agenda to assist scientists in the developing world.

I certainly did not know much about its scientific capacity building efforts, particularly those conducted on behalf of young scientists by way of its fellowships for doctorate and postdoctorate students, or its Young Affiliates programme designed to allow scientists under 40 years of age to participate in TWAS activities, most notably by attending the Academy's general meetings over the course of their five-year appointments.

These aspects of TWAS would become increasingly apparent to me in the months leading up to TWAS's 11th General Conference, held in Durban, South Africa. ASSAf was the local organizer and hosted the event, which was sponsored by South Africa's Ministry of Science and Technology.

In working with the TWAS secretariat, I learned a great deal about the Academy's broad range of programmatic initiatives that have made a difference for thousands of scientists, particularly young scientists, in the developing world.

The members are undoubtedly the core of the Academy, serving as the primary source of its prestige and influence. When I was elected to TWAS in 2011, it was a proud moment in my career – as it has been for all the other scientists who have received this honour.

But the Academy's programmatic activities play an equally important role in TWAS's success, especially in leading the way for training the next generation of scientists and for promoting South-South cooperation in science.

These programmes, in many ways, are the Academy's primary agents of change. Speak to many young scientists in the developing world and they will tell you how important the programmes have been for both them and their societies. These young scientists may aspire to TWAS membership. Yet they invariably focus their attention on the prospects for acquiring an Academy fellowship or research grant, or becoming a TWAS Young Affiliate. While older scientists (or, more accurately, older male scientists) dominate TWAS's membership, the Academy's programmatic activities are largely what are in the sights of young scientists.

Because I came to the Academy long after it had developed into a mature and esteemed organization, I did not witness – let alone participate in – the Academy's evolution from a compelling idea into a concrete reality.

In short, I did not observe first-hand the Academy's growth over the course of three decades from a small, struggling institution into a vibrant and enduring organization that now operates at the highest levels of the international science community. For those who did (and, even more so, for those who were directly involved in the Academy's development), it was unquestionably a rewarding and memorable journey, one that they rightfully tell with the passion and pride that it deserves. It is a journey from which all scientists in the developing world can learn a great deal.

Yet, for me, TWAS represents a fully realized portrait of success. The Academy has changed enormously over the past 30 years, but it has not changed much over the past five. While all of the Academy's programmes have grown tremendously in size and scope in recent years, neither their essence nor operation has changed much at all.

For better and worse, then, I see TWAS's success solely through the lens of what it does today and not what it has become over time. I think that is the way that most young scientists see it too.

The limits of conventional wisdom

I was born and raised in Durban, South Africa. I grew up thinking I would be a teacher. It was one of three potential career options that were open to young women in South Africa and many other countries in the South (and, I might add, in the North) at that time.

According to conventional wisdom, available to me through my career guidance counsellors, I could be a teacher, a nurse or a typist. While less so than in the past, the first two professions are still largely associated with women. The third, typist, has largely disappeared from the workplace, thanks to the widespread adoption of information and communication technologies – one of many examples of how science and technology fuels social change.

In many ways, when I entered the workplace, both the North and South faced a profound gender gap in science, and in most other fields. The gap was a disservice to both girls and women and to their societies. It made it difficult for 50% of the population to reach their full potential.

While the gender gap has narrowed somewhat over time, it is still apparent today.

One of the most significant differences between then and now is that virtually all nations today acknowledge that the gender gap is a challenge that must addressed. As the adage goes, you can't begin to fix a problem until you recognize that there is a one.

Changing career paths

When I began my university studies, counsellors channelled my ambitions into mathematics for the purpose of one day applying my knowledge and skills to teach in a secondary school.

I didn't particularly like pure mathematics at the university and, not surprisingly, I didn't do well either. Consequently, after two years, I switched my major fields of study to geology and physical geography. I absolutely loved geology, and we had the most inspiring teachers. But it was not considered a very suitable career for young women because the focus on mining geology and exploration required extended periods in the field.

After receiving my first degree and studying for a one-year teaching diploma, I began my teaching career in mathematics at a secondary school in Durban. It did not take me long to realize that teaching was not my destiny, and I resigned after one year. My plan was to travel abroad. I recall the school principal's final words to me that I was "making a grave mistake". Mathematics teachers, he told me, were hard to come by and in high demand.

Contrary to the principal's warnings, the year of travel turned into an opportunity for me to discover what I really wanted to do. In 1973, I returned to the University of Natal to pursue an honours and then a master's degree, specializing in my true passion: climatology. In 1975, I was appointed a lecturer and began what was to be a very fulfilling career at my *alma mater*.

I was eager to earn a doctorate degree. Yet, as was often the case at the time, there was no suitable supervisor in the university. So I began seeking opportunities to study abroad and was given sabbatical leave to pursue a doctorate degree.

I initially intended to continue my studies at the University of Waterloo in Canada. But, just before I was due to depart, I was denied a visa – not an uncommon occurrence for South African students and faculty during apartheid.

I was devastated. But alternate plans to study at the University of Virginia in the United States worked out well. I received a PhD there in environmental sciences in 1983.

I found my education at Virginia to be an exhilarating experience. The opportunity to be one of many graduate students studying in a similar field was the single most important aspect of my enthusiasm. Until then, my journey had been a lonely one. The challenges posed by expert teachers and engaged students, the competition, the ability to debate issues, experiment and take chances, all while receiving encouragement and support from the faculty, created an environment that was at once collegial and competitive – and deeply rewarding.

After receiving my doctorate degree, I was reluctant to return to South Africa. I had grown accustomed to the stimulating environment and the opportunities that had been presented to me during my doctoral studies. However, I was obliged to go back under the arrangement I had formally agreed to in exchange for receiving sabbatical leave.

Making a go of it

Upon my return, I felt very isolated. The environment had none of the excitement that I had experienced while pursuing my PhD. Since no one in my department was in the same field of study, I found it difficult to exchange ideas or even discuss my research. This was, of course, in the days before the Internet and email. My experience, I am sure, was shared by many of my contemporaries, and it still persists today in some countries, although to a lesser degree.

Having made my decision to come back to South Africa, however, I was determined to make a go of it. With hindsight, it was the best decision I could have made. I certainly would not have had the same opportunities if I had stayed in the United States. My persistence has paid off in a gratifying career.

The feelings of detachment from the mainstream that I had encountered as a young scientist began to fade when I was fortunate enough to participate in the Southern African Fire-Atmosphere Research Initiative (SAFARI-92) in the 1990s. This large international project proved to be a turning point in my career.

SAFARI-92 was a major international scientific project designed to explore land-atmosphere interactions in southern Africa, focusing largely on the impact that biomass burning was having on atmospheric chemistry in the region. The project supported a large field campaign in southern Africa, involving ground-based, aircraft and satellite measurements. I was part of a collaborative ozone group at the University of Natal responsible for obtaining measurements of ozone and meteorology in the Etosha National Park in northern Namibia.

SAFARI-92 placed me in contact with colleagues from around the world. The access it gave me to state-of-the-art information technologies and modelling techniques helped to enhance the quality of the data informing my research and broaden my research skills. The project also allowed me to examine challenges that were critical to South Africa's future, touching directly upon fundamental economic, environmental and public-health issues. I might add that it was an invaluable experience for my students too, launching many onto successful international careers.

Over the past three decades, I have been able to pursue a rewarding research agenda in the atmospheric sciences, examining such topics as air pollution and air quality management. These issues not only pose demanding research questions but are also critically important to the health and prosperity of South Africa. The same has been true of my work on climate change. I have also enjoyed my responsibilities as a teacher, especially the mentoring of master's and doctoral students.

After apartheid

As the executive officer of ASSAf for the past five years, I have become more deeply involved in examining a broad range of science policy issues that will impact the future of South Africa, most notably, education, sustainable development and health.

Throughout my career, I have both observed and participated in the historic changes that have taken place in my home country in the aftermath of apartheid and the rise of a multiracial democratic society.

The ostracism faced during apartheid affected all South Africans. The scientific community, while often holding a privileged position internationally by being able to maintain contacts abroad, was not exempt from the world's condemnation of our national policies. I recall struggling to obtain a visa to attend a conference in Australia in the late 1980s. I likely would have been denied it except for the intervention of the International Council for Science (ICSU).

The situation was quickly reversed with the fall of apartheid in 1994. In fact, the mid- and late-1990s proved to be a period of expansive opportunities for South African scientists. Many colleagues from around the world wanted to collaborate with us. At the same time, governments and funding agencies in the United States and Europe were seeking to extend their support to the 'new' South Africa, which was eagerly seeking to establish a multiracial democracy under the leadership of its iconic president, Nelson Mandela.

As the boycotts and travel restrictions were lifted, new opportunities opened up for South African scientists to work on international projects and nurture contacts abroad. One of the reasons that South Africa's scientific community was able to rejoin the international scientific community so quickly is that the nation enjoyed deep pockets of scientific excellence that rivalled the best in world in such fields as archaeology, astronomy, infectious diseases, mineralogy, and plant and animal science.

As a result, in certain disciplines, South Africa did not have to build scientific capacity and excellence to forge cooperative arrangements with partners in the scientifically most advanced countries. Scientists simply had to be given the opportunity to share their expertise. Moreover, a number of universities in South Africa, those that had served the white minority population, were the best in Africa and among the best in the developing world.

Despite its obvious strengths, science in South Africa has not been able to fully escape the legacy of apartheid. This legacy manifests itself in a number of ways.

Stubborn educational inequities

Critical issues in national science policies and many other fields involve not only capacity and excellence, but also equity and access. For South Africa, in light of its history, issues of equity and access carry particular relevance. At the level of tertiary education, this means ensuring black students are adequately prepared for the country's best institutions. It also means improving the quality of education in historically disadvantaged black universities that were woefully underfunded during apartheid.

The problem extends beyond higher education to South Africa's primary and secondary school systems. Nearly two decades after the end of apartheid, we have yet to improve the quality of our national education system. Funding has increased but educational outcomes have not, especially among students in black communities.

Having excellent universities that remain beyond the reach of the majority of students – because students are ill-prepared for the rigours of a university education – cannot continue if South Africa hopes to create a society that provides equal opportunity for all citizens.

The nation is keenly aware of the problem and has discussed it at length for years. But progress, let alone a solution, has yet to materialize. Until primary and secondary school education improves significantly, universities in South Africa will continually be handicapped in their efforts to balance excellence and access in their student enrolments.

That is why ASSAf has devoted a great deal of time and effort to examining ways to improve education, particularly science education, in both primary and secondary schools. The academy believes this is a key building block for achieving greater scientific capacity across the country and for ensuring that science is an instrument used to improve the lives of all South Africans.

New networks of cooperation

The end of apartheid created opportunities for researchers in South Africa to partner with colleagues across the African continent for the first time in decades. It also opened the doors of South Africa's universities to students from other African countries. These developments have proven beneficial to the entire continent.

In fact, scientific collaboration among African nations now constitutes a regional dimension of the broader trend in South-South collaboration in science, which has manifested itself in an increasing number of programmes. This, too, has proven beneficial to the entire continent. South Africa, for example, has forged bilateral agreements for scientific collaboration with many African countries, including Algeria, Angola, Kenya, Namibia, Mozambique and Tanzania.

At the international level, South Africa is home to the African Laser Centre (ALC) and the African Institute of Mathematical Sciences (AIMS). There's also the Cape Town component of the International Centre for Genetic Engineering and Biotechnology (ICGEB) at the University of Cape Town. ICGEB's headquarters is in Trieste, Italy, and there is another component in New Delhi. The Cape Town component focuses on research initiatives of special importance to Africa – for example, infectious diseases and plant biotechnology. It also has a strong training programme in addition to its research agenda.

Other examples of international collaboration include the postgraduate training fellowship programme for women scientists from sub-Saharan Africa and Least Developed Countries (LDCs), which is sponsored by the Swedish International Development Cooperation Agency (Sida) and managed by the Organization for Women in Science for the Developing World (OWSD). The programme has played an important role in helping women scientists in Africa earn advanced degrees. Many of the recipients from sub-Saharan Africa, who under the terms of the fellowships are required to return home upon receiving their degrees, are being educated in universities in South Africa.

Meanwhile, the African Science Academy Development Initiative (ASADI), which is sponsored by the Bill and Melinda Gates Foundation and managed by the US National Academy of Sciences, has helped to link African science academies together in joint research projects. Historically, scientific collaboration in South Africa has taken place among individuals. ASADI hopes to lay a foundation for stronger institutional collaborations.

The last two initiatives are closely associated with TWAS, due to its ties to OWSD and IAP, the global network of science academies. The secretariats of both these organizations are located in Trieste and TWAS plays an important role in their administration.

The power of stargazing

The most significant international scientific project is the Square Kilometre Array (SKA) radio telescope project. SKA will be the world's largest and most sensitive radio telescope. There will be two core sites – one in South Africa and one in Australia. A broad network of telescope dishes and aperture array stations will be built across southern Africa – in Botswana, Kenya, Madagascar, Mauritius, Mozambique, Namibia and Zambia. The final design phase for SKA began in autumn 2013. Construction is scheduled to start in 2017, and the project is expected to be fully operational in 2024.

South Africa enjoys a long and accomplished history in astronomy. The first astronomical observatory dates back to the colonial period in 1685. Today, South Africa boasts one of the world's most expansive networks of telescopes and observatories, which includes the South African Astronomical Observatory, the South African Large Telescope and MeerKAT, currently under construction.

SKA, which will engage thousands of scientists and 18 countries, promises to create broad opportunities for international collaboration in physics as part of a quest to answer fundamental questions in physics, astrophysics, cosmology and astrobiology. Equally important, it promises to help forge closer links among physicists both globally and across southern Africa.

As the sampling of initiatives cited above shows, scientific cooperation and collaboration between South Africa and other African countries is growing. South African scientists, nevertheless, continue to prefer forging partnerships with colleagues in Europe and the United States.

Such partnerships have historic roots. But they also reflect a universal desire on the part of scientists – indeed people in general – to seek the best opportunities that are available to them so that they can maximize their potential.

This universal desire has placed South Africa in a unique position. On the one hand, South Africa has become the preferred destination for students and professionals of other African countries. On the other hand, South Africa also has served as a way station for African students and professionals, who ultimately seek educational and job opportunities in the North. As a consequence of these dual trends, South Africa has been both a beneficiary and a victim of brain drain.

In a similar vein, South Africa has been included as one of the BRICS. This group of countries – Brazil, Russia, India, China and South Africa – has been among the fastest growing countries in the world. Economists estimated that, collectively, the BRICS have accounted for nearly one-third of the world's economic growth since 2000 and more than half of the growth since 2008.

When it comes to science, by some measures, South Africa is worthy of inclusion in this group. For example, according to Shanghai Jiao Tong University's 2013 ranking of the top 500 universities in the world, there are three universities in South Africa in this prestigious listing: the University of Cape Town, the University of the Witwatersrand and the University of KwaZulu-Natal. Brazil, in contrast, has six, Russia two and India one. As reflection of its rapid increase in scientific capacity and excellence, China now has 28 universities in the top 500 ranking. So, on this score, South Africa's best universities line up well with most of the other BRICS countries.

However, when it comes to the percentage of young people attending universities, South Africa falls behind the other BRICS nations. According to the *World Economic Forum's Global Competitiveness Report 2013-2014*, 75.9% of Russia's tertiary age population is enrolled in tertiary education (among the highest percentages in the world). In China, the proportion is 26.8%; in Brazil, 25.6%; in India, 17.9%; and in South Africa, 15.4%. If we also consider the low percentage of black students enrolled in South Africa's best universities, then the country is clearly not keeping pace with other BRICS countries in terms of graduate and postgraduate study. This has serious implications for South Africa's ability to remain competitive, not only with the other BRICS countries but also with all countries.

The challenge of being heard

All of these issues have been central to my work since my appointment as the executive officer of ASSAf in 2008.

After being employed for some three decades as a researcher and teacher at the University of KwaZulu-Natal, I welcomed a new challenge.

I had thoroughly enjoyed my career as teacher and researcher. I had found my participation in several high-profile research projects to be extremely gratifying and I was pleased to have published a number of widely read academic articles in the field of atmospheric science. I also found teaching to be extremely rewarding. As I mentioned above, I believe educating the next generation of scientists in South Africa to be of critical importance to its future, and I was proud to be have contributed to this effort.

Despite all this, I felt it was time to move on. In 2008, I applied and was fortunate enough to be appointed ASSAf's executive officer.

As part of the arrangements for the lead-up to the TWAS conference in Durban, South Africa, in 2009, ASSAf prepared a book on the state of the basic sciences in South Africa.

Although the book did not offer recommendations on how to improve science, it did provide comprehensive benchmarks for assessing the breadth and depth of research taking place in conventional fields of study – for example, biology, chemistry, earth sciences and physics. In this sense, it complemented other reports that have been produced about the state of science in South Africa. Written by scientists for policymakers, the goal was to provide a detailed portrait of science in South Africa presented by those directly engaged in research.

The book has been favourably received. We would like to publish an updated edition in the future and even add chapters containing similar information about such multidisciplinary and frontier fields of study as biomedicine and nanotechnology.

In addition, we recently completed a report on science, technology and innovation in South Africa. As stated earlier, South Africa faces major challenges in ensuring that its workforce will have the education and skills that it needs to succeed in the 21st century. South Africa is simply producing too few PhDs in science to meet the demands of the future, especially in light of the competition it faces from countries that are rapidly improving their scientific capacity.

Challenges concerning poverty, unemployment, housing, health and inequality dominate the political debate in South Africa. Because of these daunting problems, issues related to scientific and research capacity are often placed too far down the list of national priorities. Such competing concerns have sometimes made it difficult for science advocates to be heard.

Scientists thus have an important role to play in convincing both public officials and the public at large of the value of science to society. ASSAf views this issue as one of its most important responsibilities.

TWAS's diversity deficit

The 2009 TWAS conference proved a significant event for both TWAS and ASSAf. The minister of science, Naledi Pandor, spoke passionately about the importance of science at the conference. The full range of scientific research activities taking place in South Africa was put on display for scientists from around the world to see. Young scientists from South Africa were able to interact with their colleagues from developing countries and emerging economies. And TWAS officials were granted a special meeting with the president of South Africa, Jacob Zuma, to discuss broad issues related to science and development.

All in all, the conference raised the profile of science in South Africa and helped South Africa's scientific community forge closer ties with colleagues from other countries.

Since then, I have been increasingly involved in TWAS activities, focusing on two major issues: the election of new TWAS members and assessing issues of gender equality in the Academy.

Following my election to TWAS in 2011, I became a member of the TWAS advisory membership committee in astronomy, space and earth sciences.

Seeking to identify the most qualified candidates for membership is a critical responsibility. The committee, not surprisingly, takes its responsibilities seriously. The procedures are exacting and the review process meticulous.

Having said that, TWAS clearly faces a tough challenge in seeking to attain greater geographic distribution among its membership. Historically, Academy membership has been highly skewed towards scientists in large developing countries with rapidly increasing scientific capacity – most notably, China, India and Brazil. Recent outcomes in the election process indicate that this trend remains stubbornly in place. In fact, the growing number of excellent scientists being produced in these countries means that their candidate pools will likely deepen in the years ahead.

I do not favour turning to "affirmative action" to address this issue. But I do believe the Academy should take strong measures to expand membership representation from countries that historically have been under-represented.

First, I think TWAS members from under-represented countries must put forth a larger number of candidates for membership since expanding the pool of candidates from these countries would increase their chances for election. In 2013, for example, there was just one candidate from South Africa across all disciplines. Clearly, we can – and must – do better.

Second, I believe that the TWAS advisory membership committee should give top-ranking candidates from scientifically lagging countries some special consideration in the vetting process. Competition among candidates is fierce. Differences between the best candidates in terms of scientific abilities and accomplishments are often exceedingly narrow. In those cases where the high level of excellence among the finalists is unquestioned and virtually the same, I believe the Academy should exercise its judgment and lean towards the candidate from a scientifically lagging country.

Third, the Academy should continue to strengthen its fellowship and research grant programmes for young scientists, especially those from scientifically lagging countries.

Over the long term, there is no better way to broaden the potential pool of candidates for TWAS membership, grants and prizes than to provide ample high-level educational and training opportunities for young scientists in all countries. Nurturing scientific talent everywhere may ultimately be the most effective solution to the puzzling Academy challenge of skewed representation among its membership.

Overcoming gender inequality

The Academy faces a similar issue when it comes to the problem of gender inequality among its membership. It is not an unfair caricature when people describe TWAS (or, for that matter, other science academies) as an old men's club. It is a reflection of reality.

The Academy has slowly increased its percentage of women members, both thanks to its own proactive efforts and the increasing number of women seeking careers in science. But the problem continues. It is especially acute for women scientists from scientifically lagging countries. These countries simply have too few women scientists to provide a reasonable number of candidates.

The TWAS gender advisory board, of which I am a member, has discussed the gender imbalance issue at length. One recommendation that the board plans to make to the TWAS Council is for the Academy to prepare an annual report detailing the level of involvement of women in various activities. For example, the report would list the number of women who are members, who have served on committees, who have received research grants and awards, who have joined conference-organizing committees, and who have participated in conferences as moderators or speakers.

In effect, these annual reports, designed to raise awareness about gender gap issues, would describe in detail the state of women in the Academy. Hopefully, the reports would provide the impetus to devote even more attention and resources to this critical issue.

The Academy should be commended not only for taking gender issues seriously but also for the progress it has achieved. Nevertheless, the Academy's commitment has depended too much on the goodwill of the Council. I would like to see a formal strategy, documented in writing, which would provide a systematic blueprint for examining and addressing this critical challenge. Once benchmark statistics have been established, follow-up assessments could then be prepared to gauge the progress – or lack of progress – that has been made.

Continual self-assessment

During the brief time that I have been associated with TWAS, I have learned a great deal about the value and importance that scientists in the developing world have placed in the Academy. Equally important, I have learned about the Academy's worthy efforts on behalf of young scientists.

TWAS is certainly recognized as a significant voice for science in the South and an important agent of change.

However, in an era of historic change in global science, agents of reform like TWAS have no choice but to change continually. It is the price an institution must pay to remain pertinent. Indeed, successful institutions that are willing to engage in continual self-assessment and reform receive the most valuable payoff of all – a future even brighter than the past.

For generations to come

Maria Corazon A. De Ungria

Maria Corazon A. De Ungria, one of the first TWAS Young Affiliates, heads the DNA Analysis Laboratory in the Natural Sciences Research Institute at the University of Philippines, Diliman Campus.

De Ungria attended a local primary school. She then went to the Philippine Science High School, a government-sponsored secondary school designed for scientifically inclined and academically gifted students, graduating in 1985. Following two years of study at the University of the Philippines, she travelled to Australia, earning a Bachelor of Science degree in biology from Macquarie University in 1993 and a doctorate degree in microbiology from the University of New South Wales in 1999.

De Ungria lived in Australia for more than a decade, returning to the Philippines in December 1998. Two months later, she was appointed head of the DNA Analysis Laboratory at the University of Philippines.

Over the past 15 years, De Ungria has helped elevate the laboratory into a well-respected research centre and a valuable national asset that collaborates with a broad range of institutions in the Philippines.

Under De Ungria's guidance, the laboratory has been a technical adviser to the Supreme Court's efforts to formulate rules for the application of DNA evidence in courtroom proceedings. In addition, the laboratory has promoted the use of DNA evidence in paternity and sexual abuse cases and has led a successful campaign to utilize DNA evidence in post-conviction capital cases. It has also served as a centre of research and education dedicated to exploring the genetic ancestry of the Philippines' diverse population.

In addition to her responsibilities at the DNA Analysis Laboratory, De Ungria directs the programme on forensics and ethnicity at the Philip-

pine Genome Center at the University of the Philippines and is a lecturer in the science and society programme. She has published more than 30 scientific articles and has also written news and magazine articles for the popular press.

De Ungria has been the recipient of numerous awards and grants. Her most recent honours include the Outstanding Research and Development Award for Applied Research from the Philippine Department of Science and Technology; For Women in Science National Fellow, sponsored by L'Oreal and the United Nations Educational, Scientific and Cultural Organization (UNESCO); and travel grants from the European Commission's Erasmus Mundus MAHEVA (Man, Health, Environment, biodiVersity) in Asia Programme, the Royal Society of London and National Science Foundation of China to visit laboratories in Spain, the UK and China.

In 2006, TWAS and the Philippines' National Academy of Science and Technology presented De Ungria with the Outstanding Young Scientist Award in Biology. In 2007, she was named one of five young scientists in the inaugural class of TWAS Young Affiliates from East and Southeast Asia.

In the following essay, De Ungria describes the path that led her to a successful career in science. She then outlines her current research agenda and outreach efforts. She concludes by explaining why she believes that the story of Abdus Salam and the creation of TWAS remain relevant today to young scientists in the developing world.

***

I first became aware of TWAS as a consequence of reading the material that the Academy periodically mailed to the DNA Analysis Laboratory at the University of Philippines Natural Sciences Research Institute.

The year was 1999. I had just returned home after spending more than a decade abroad, earning advanced university degrees in Australia. I had been looking for a suitable job for just a few weeks before being appointed head of the laboratory.

Funding was scarce. Consequently, one of my main responsibilities was to actively search for money. Since Internet access at the time was limited and slow, snail-mail continued to be a major source of information. Overseeing the daily mail call gave me a window to the outside world. Saturnina C. Halos, the former head of the laboratory who is now a consultant, immediately took a special interest in my career, providing me with sage mentoring advice. Halos viewed this task not as a clerical exercise but as an important gatekeeping responsibility. I couldn't have agreed more.

Imagine that: learning about TWAS via land mail – printed brochures, pamphlets and flyers – that arrive weeks after they were sent. That's likely not the way young researchers learn about the Academy today in the age of the Internet and social media.

The array of programmes and activities sponsored by TWAS was impressive. But I wasn't simply interested in browsing through the Academy's promotional material. One of my primary goals was to learn about funding opportunities.

I discovered from the mailings that the Academy supported visiting lectureships and purchases of equipment and supplies. The laboratory applied for a small grant in each of these funding categories and was fortunate to be awarded both.

But before I discuss how these grants helped to move the laboratory and my career forward, I think it might be useful to outline how I had reached this point.

Education matters

I was born in 1968 into a middle-class family in Manila. My dad was a chemical engineer who worked for Meralco, the country's leading energy and power-distribution company. My mom was a dentist who had given up her career to be a full-time housewife and to care for my three brothers and me. Societal priorities in the Philippines and throughout much of the world in those days meant that women often relinquished their careers (if they had one) to raise their children. My mother did not have to work to augment my father's income. Neither did my brothers or I. By our good fortune, we could devote all of our time to our studies, which our parents encouraged.

I received an excellent education. At age 11, I was selected to attend the Philippine Science High School after taking two qualifying exams and being interviewed by school officials. This marked the first, but by no means the last, time that I would be asked to answer questions about myself as part of a selection process.

The Philippine Science High School is a special government school for students who have displayed scholastic excellence in primary school and a strong inclination toward the sciences. Because students do not pay tuition and receive a monthly stipend from the government, they are often referred to as "scholars of the state".

Transferring to the Philippine Science High School meant leaving old friends and finding new ones among the 240 students who were enrolled in my class. Students came from diverse cultural and economic backgrounds. Some lived in farms and villages. Others resided in gated subdivisions. The students represented a microcosm of Philippine society – a reflection of the country's demographic diversity and limitless potential as embodied in its youthful population.

The Philippine Science High School provided an exciting and rewarding educational experience. The school both stimulated my intellect and broadened my perspective on the world. I have been forever grateful for this experience. It helped teach me the value of hard work, resilience and open-mindedness.

Some 30 years later, I remain an active alumna and a vocal supporter of the school. At the time, it was a one-of-a-kind facility. As a testament to its success, the government has since built a number of regional schools, making this type of educational and learning environment accessible to more students. As a result, the Philippine Science High School is now part of a high-achiever scholastic system benefiting several thousand students each year.

In primary school, I had displayed an aptitude for science and literature. My secondary school, however, was dedicated to educating future scientists. Students were required to take specialized subjects in physics, chemistry and biology just a year after admission, unlike students in other schools who were assigned courses in science only in their final two years. In addition, in the third and fourth years, we had to take "research" as a discrete subject and were expected to select our own research topic and carry out experiments with little supervision from our teachers. We engaged in every facet of the project – from devising a challenging research question, to securing funding (my parents were my first funding source), to doing the research, to recording and interpreting the results.

The hands-on learning experience allowed for a great deal of independence. It was something that I thoroughly enjoyed, both for the challenges it afforded and also for the sense of empowerment it gave me.

The experience hooked me on science not just as an exercise to fulfil a course requirement for graduation, but also as a personal preference. As I look back today, I think that my secondary school learning experience provides invaluable lessons on how to encourage young people to enjoy science both at school and, potentially, as the basis of an interesting and rewarding career.

True ambitions

Following graduation, I enrolled at the University of the Philippines, where I expected to major in zoology as a pathway to becoming a medical doctor. Studying zoology was then viewed as a prerequisite for applying to medical school.

However, after two years of study, I realized that my true ambition was to become a scientist. Doctors belong to one of the world's most noble professions. But I knew that laboratory research, not patient care, was my calling. I also believed that my chances to have a broader and more lasting impact on a society would be enhanced if I pursued a career in research where I would be at the forefront of exciting discoveries that could potentially help improve the lives of the Filipino people.

So I left the Philippines in 1988 to study in Australia. I didn't anticipate that I would be gone from home for more than a decade – first to earn a bachelor of science degree in biology (with honours) from Macquarie University in 1993 and then to earn a doctorate degree in microbiology from the University of New South Wales in 1999.

Upon completing my degree, I was tempted to remain abroad. This feeling was reinforced by several attractive offers to pursue postdoctoral studies in Australia, the US and the UK. However, I chose to return home.

The reason for this decision was simple. I profoundly missed my family whom I had visited only a few times since leaving for Australia. The cost of travel was expensive and, during summers, I worked full-time to add to my finances.

To be absent from my family for more than a decade came at a great personal sacrifice. It was not something I wanted to continue – or even worse, to make permanent. I returned to the Philippines just ten days after submitting my thesis.

I also felt a responsibility to give something back to my country. The government – and thus taxpayers – had covered the cost of large portions of my education since secondary school, helping me to receive excellent instruction in the Philippines and abroad. I was thankful to be the recipient of such generosity and I strongly believed that I had a duty to apply my knowledge and skills to address critical problems in my own country, in the hope of improving the lives of all Filipinos. Adrian Lee, a professor of medical microbiology at the University of South Wales, who had served as my advisor, admired my patriotic ideals and encouraged me to go back home.

Nevertheless, upon returning I still harboured doubts about whether I could succeed. Launching a scientific career in the Philippines after having been away for so long posed daunting challenges.

In Australia, I had become accustomed to a research environment where the tools of the trade were readily available. First-rate equipment, ample laboratory space and sufficient funds for supplies meant that researchers could concentrate their time on the projects in which they were involved. Large networks of faculty and students in comparable fields of study, moreover, created a collegial atmosphere that both nurtured and challenged one's efforts.

Determination, and a little luck

Upon returning home, I was prepared to join the academic community and devote my time to teaching. I was not sure, however, whether I wanted to establish my own laboratory. In fact, I was more inclined to join an existing laboratory, which would allow me to sidestep the difficult task of building my own facility.

Fifteen years ago, many laboratories in the Philippines were small and woefully underfunded. It was especially difficult for new PhDs to find the necessary funds to build a laboratory, which would entail purchasing equipment and hiring qualified staff.

Yet, at the same time, it was also difficult to find a suitable postdoctoral position in a well-funded laboratory. Within weeks of coming home, I had to confront the reality that the challenges I faced in seeking suitable employment were far greater than I had ever imagined.

Then I got lucky. A few days after celebrating New Year's in 1999, I decided to board a "jeepney", a common form of public transportation in the Philippines, to return to the University of the Philippines to see how things had changed over the years.

Classes were not yet in full swing following the holiday break and I found myself walking around the campus virtually alone. Then I came across the Natural Sciences Research Institute, which I had never ventured into during my two years at the university. I approached the guard to ask permission to look around. Because I was not part of the institute, the guard alerted Saturnina C. Halos, former head of the DNA Analysis Laboratory, who happened to be visiting there. (Since she had become a consultant, she did not go to the laboratory regularly.) We chatted about my background and my desire to begin working as quickly as possible.

That chance meeting in January 1999 became one of the most important encounters in my career. Within a month, I was appointed head of the DNA Analysis Laboratory.

In some sense, my ambition-driven luck was created by choice, not chance. It was a reflection of my personality to try to make things happen rather than wait for them to happen – a cherished characteristic rooted in the values that my parents had placed on hard work and education. It was also a reflection of my love for learning and discovery that my teachers and mentors had nurtured.

I also believe, however, that chance plays an important role in a person's life. In my view, a clear factor in success is one's ability to maximize the opportunities that chance brings.

By virtue of a fortuitous encounter, Halos would become both my hiring agent and mentor. And by virtue of the early assignments she gave me, she was the person most responsible for my discovery of TWAS.

Research and impact

The appointment allowed me to begin work in a fully functioning laboratory. Yes, the laboratory was small – there were only two positions other than the head. Yes, the equipment could use an upgrade, if not a complete overhaul. And, yes, the laboratory would benefit greatly from forging closer ties to the scientific community, especially the international scientific community.

But no, I would not have to find funds to build or renovate the infrastructure. I could spend my time doing research and finding funds for new projects.

Much of the initial work of establishing an operational laboratory had already been done. My primary task was to nurture this small laboratory into a thriving research facility with sufficient influence to effect societal change through excellent science.

As an added bonus, the laboratory was also conducting interesting work in a cutting-edge field – forensic DNA technology – that was closely allied to my studies in micro- and molecular biology. This meant there would be opportunities for expanding the laboratory's activities and that I was equipped, by virtue of my education and training, to explore what those opportunities might be.

Besides conducting basic genetic research, laboratory staff also provided DNA testing for determining biological parentage and paternity for child support and inheritance claims, as well as for incidences of gross negligence that had led to infant switching immediately following birth.

Such activities supplied the laboratory with additional revenues beyond its support from the government. But even more importantly, it enabled the laboratory to engage in issues of critical importance to society.

DNA testing supplies valuable insights into civil and criminal investigations that cannot be acquired through conventional means. The laboratory firmly believes that this is an important community (indeed national) service – a view shared by government officials and the public at large.

Such efforts conform to my belief that the laboratory – and, by extension the university – should conduct research not just for the sake of publishing in academic journals, but also for the sake of applying that knowledge to improve communities and the lives of people.

In line with these goals, we have since expanded our DNA techniques to identify perpetrators in civil and criminal cases involving sexual abuse; investigate whether individuals have been falsely tried and convicted, especially in capital cases; and ease the identification of those who have tragically lost their lives in natural disasters.

These activities do not diminish the central role that basic research plays in the laboratory's agenda. During my tenure as head of the laboratory, one of my primary responsibilities has been to secure funding for such research activities. As I mentioned earlier, this was how I became acquainted with and interested in TWAS.

Deeper, broader involvement

As any observer will tell you, the university has three critical functions that broadly define its contributions to society: teaching, research and community service. The proportion of time devoted to each may vary from institution to institution and from time to time. But all three are critical responsibilities that institutions of higher education must embrace if they are to be viewed as effective.

The work of university scientists, I firmly believe, must not be confined solely to the scientific community but must also resonate in policy circles. Such wider fields of involvement, which have been at the centre of the laboratory's agenda since its inception, speak to responsibilities that I have sought to expand ever since I was named the head of the facility.

For example, the laboratory served as a technical adviser to the Supreme Court in its effort to formulate national rules for the use of DNA evidence in legal proceedings. In promulgating the Rule on DNA Evidence in 2007, the court found applications of such evidence to be relevant and valuable. It also prescribed guidelines for the acquisition and preservation of DNA evidence.

These procedures are essential for ensuring the validity and admissibility of DNA evidence and for protecting individual privacy rights. The law was informed – indeed shaped – in part by the expert data and information that the laboratory brought to bear on the discussions that led to the promulgation of the rule.

After conducting its own review of capital cases between 1993 and 2006, in which it determined that more than 75% of the convictions might have been erroneous, the Supreme Court subsequently included a section on post-conviction DNA testing in the Rule on DNA Evidence.

This section prescribes that cases, which had undergone final review, may be re-opened if relevant biological samples have not yet been tested. Notably, in 2006, the death penalty law was repealed in the Philippines because of public concerns that innocent people had been wrongfully convicted. The laboratory helped to raise public awareness by contributing DNA evidence in several cases in which the accused were acquitted. To further aid in judicial reform and assist individuals serving sentences who claim to be innocent, the laboratory helped to form the Innocence Project Philippines Network. The network, officially launched in December 2012, also includes law schools and legal aid clinics.

By serving to uphold the human rights, especially the rights of individuals who may have been wrongfully convicted, we believe that the use of DNA evidence has helped to raise public confidence in the nation's judicial system.

Embracing excellence

All of these efforts require financial resources. They also require the ability to conduct research to ensure that the quest for social justice is based on a sturdy foundation of science-based evidence. Simply stated, striving for excellence in science is at the core of what we do.

That is where our relationship with TWAS has come into play. The laboratory was awarded its first TWAS grant in 2001. Funds allowed us to sponsor the visit of Wing Kam Fung, a distinguished professor of statistics and actuarial science at the University of Hong Kong, for a series of lectures. The grant enabled staff and students to gain valuable insights into forensic biology. It also helped to open the institution's doors to scientific collaboration with institutions beyond our nation's borders.

The second grant from the Academy, awarded in 2002, provided money for the purchase of molecular reagents for processing biological samples from male volunteers for the purpose of establishing a Y-chromosomal

DNA database. The laboratory analysed the samples to determine the genetic variation of selected Y-chromosomal DNA markers in the Philippine population. These markers provide a powerful tool for male identification for use in forensic cases and human evolutionary studies.

Both grants were small. Yet their value extended far beyond their monetary worth. They gave the laboratory a modicum of international prestige that it could point to when seeking additional support, especially from governmental sources in the Philippines. They instilled a sense of pride among the staff and students. They encouraged us to seek funding from other organizations. And they made us believe that the world, at least the scientific world, was paying attention to our work and considered it of value. All in all, I believe that the grants have proven to be an excellent investment for both the laboratory and the Academy.

TWAS honours

In 2006, I was named an Outstanding Young Scientist, an honour bestowed by TWAS and the National Academy of Science and Technology in the Philippines. One year later, I received an email from TWAS that I had been named a TWAS Young Affiliate.

I was truly surprised to receive the Outstanding Young Scientist Award, and I didn't even know what a TWAS Affiliate was, let alone the benefits and responsibilities that would accompany the appointment. Yet it was deeply gratifying to have received this recognition by virtue of the organizations that were serving as the award sponsors. Each honour has proven instrumental in raising my profile in the scientific community, especially in the Philippines.

I soon learned that Affiliates were young scientists from the developing world who had shown promise and who had been chosen through a meticulous vetting process. I also learned that the TWAS Young Affiliates programme was a new initiative and that I would be a member of the inaugural class, which was comprised of 25 members.

Appointees, who are awarded five-year terms, are entitled to participate in TWAS activities, most notably by attending the Academy's general meetings as well as conferences organized by the Academy's regional offices.

Through such activities, Affiliates are placed in contact with other young scientists from the developing world, providing opportunities to nurture a global network of our peers. The programme resembles a junior academy for young scientists, which is intended to give a sustained boost to promising careers.

Being a TWAS Young Affiliate has increased my visibility both in the Philippines and internationally. The announcement was publicized in the scientific community (and to a lesser degree in the popular press) across the country. Messages of congratulations reflected that others appreciated the value of being selected.

I am delighted that since my appointment, three additional Filipino scientists have been named Young Affiliates (in 2009, 2010 and 2013) and that two other scientists from the Philippines have won TWAS regional prizes for the public understanding and popularization of science (in 2010 and 2013).

Raising the profile of TWAS

As a TWAS Young Affiliate, I was also given an opportunity to travel abroad, including to Trieste. Only then was I able to learn about the full scope of TWAS's activities and the Academy's rich history in helping to build scientific capacity in the developing world. Only then, moreover, was I able to learn about Abdus Salam's contributions to science both as a preeminent scientist and humanitarian.

During my trip to Trieste, I saw a documentary video on the life of Salam entitled *Abdus Salam: The Dream of Symmetry*. I secured a copy of the film and subsequently sent it to Christopher and Maria Victoria Bernido, a husband-and-wife team responsible for the "Learning Physics as One Nation" project. The initiative seeks to improve physics instruction in 200 secondary schools in the Philippines, many in remote areas. I thought it was important to share the story of Salam with them, which I believe could help them inspire their students. I view such efforts as a crucial responsibility of all Affiliates and, in fact, of all scientists who have been involved with TWAS.

Since my appointment as a Young Affiliate, I have sought to raise the profile of TWAS among both scientists and scientific institutions in the Philippines. At the same time, I have also sought to raise the profile of Filipino scientists in TWAS.

Such efforts, I believe, could be accelerated if more vigorous communication were to take place at higher levels of administration and management. That is why I think it is important to foster greater information exchanges between the president of the Philippine National Academy of Science and Technology and the leaders of TWAS, especially concerning the current state of science in the Philippines and the most critical needs faced by scientists here.

Across generations

For young scientists from the developing world, TWAS is an important institution. It serves as a valuable source of funding for research and a key networking hub for colleagues and collaborators. Having ties to TWAS – through its grants and awards programmes or as a Young Affiliate, or both – generates multiple benefits through the support and prestige that are bestowed.

These ties help to build one's *curriculum vitae* and boost confidence, enhancing the chances for a successful career. On a more personal level, they help to nurture contacts that create friendships and collaborations that could last a lifetime.

Young scientists in the developing world, of course, may aspire to become a TWAS member. This is a dream undoubtedly worth embracing. Nevertheless, it is the Academy's more immediate benefits that most often attract their attention.

In my estimation, that is the way it should be. If young scientists are devoting the majority of their time and energy to their careers in a highly competitive environment marked by both promise and pitfalls, then science in the developing world must be approaching the international norm. The challenges for young scientists in the South are becoming personal, not global, much like the experience of young scientists in the developed world. In my mind, this is another indication that the North-South gap in science is closing.

Yet, I would be quick to add that the history of TWAS – and the enormous challenges that Abdus Salam and his colleagues encountered in seeking to establish and sustain the institution – remains an important story. It should command the attention of current and future generations of scientists in the developing world. The details of the challenges that were embodied in this effort will likely become lost to history unless young scientists feel compelled to learn about them.

That is why I strongly believe that young scientists should spare some time to learn about the life of Abdus Salam and others in TWAS who have been key agents of change both in their own countries and on a global scale.

"On the shoulders of giants"

As Sir Isaac Newton once observed: "If I have seen further, it is only by standing on the shoulders of giants." He was referring to advances in scientific research. But I believe that these sentiments apply equally well to advances in science policies. Greater funding, improved research facilities and broader international collaboration are the lifeblood of scientific excellence.

While the unprecedented growth in science in the South over the past three decades may seem rapid in hindsight, progress in most countries – including the Philippines – has taken place not in leaps and bounds, but step by step. Much like advances in research, moreover, it has been a reiterative process that has often drawn on the vision and strength of others. All of us, in some sense, benefit from the work and accomplishments of those who came before us who were determined to improve the research environment in developing countries.

My generation owes a great deal of gratitude to Abdus Salam and his colleagues. They engaged in a full range of activities – from superlative research to endless advocacy – in ways that helped to set the stage for the extraordinary growth of science in the developing world that has taken place over the past three decades.

Their story is an inspiring one that needs to be told time and again. It will help us to better understand the journey that has led us to where we are today and will also shed light on where we have to go from here.

Today's generation of scientists do indeed stand on the shoulders of giants – scientists like Abdus Salam and all those who played such a critical role in the creation and development of TWAS. If we do our job with equal skill and dedication, scientists of the next generation and for generations to come will stand on our shoulders as well.

Photo credits

Juliana Bressan 74
Michelle Burgos / Forbes Mexico 114
Chinese Academy of Sciences viii
ICTP Photo Archives 50
Nick Jackson / Blackett Laboratory, Imperial College, London 14
Edward Lempinen / TWAS 178, 231
Gláucia Rodrigues / Pesquisa, FAPESP 28
Massimo Silvano / TWAS 90, 128, 235
TWAS Photo Archives 160
Photo courtesy of the author 144, 194, 212

Front cover:
The Banaue Rice Terraces, located on the Philippine island of Luzon, have been designated a National Cultural Treasure of the Philippines and a UNESCO World Heritage Site. Adi.simionov/Wikimedia Commons (CC BY-SA 3.0)

Back cover:
An overcrowded hilltop barrio in Mexico City. Roberto Hernández/Flickr (CC BY-ND 2.0)

Photo credits

Acknowledgements

A Voice for Science in the South *was commissioned by TWAS as part of an ongoing effort to record the Academy's past and explore its future. Numerous people have helped to make it possible.*

Daniel Schaffer, who served for 15 years as the TWAS information officer, provided impressive energy and historical knowledge in bringing this volume to life. Members of the TWAS Council have also enthusiastically supported this initiative, viewing it as a valuable addition to the TWAS story.

Special thanks to the Academy's past presidents, José Vargas, C.N.R. Rao and Jacob Palis, and to current President Bai Chunli, all of whom agreed to explore TWAS's success and challenges in these pages. We are deeply indebted to Mohamed H.A. Hassan, the Academy's long-time executive director, who was – as always – an invaluable source of knowledge and insight. Indeed, a sincere thanks to all of the contributors; they have taken time from their schedules to examine their involvement with TWAS and, more generally, to speak about a time of dramatic change both in their own countries and across the globe.

My heartfelt thanks to the TWAS secretariat, whose consistent good work supported this volume in ways both direct and indirect. Thanks especially to Public Information Officer Edward Lempinen and to the PIO staff: Gisela Isten, who has been with TWAS since its earliest days, Sean Treacy and Cristina Serra. This book owes much to their expertise and creativity. Thanks as well to Rado Jagodic at Studio Link in Trieste, Italy, who has lent his considerable talents to the design of this book.

Finally, I would like to express appreciation to the members of TWAS, now more than 1,100 strong, who have made this organization such a remarkable institution – one that deserves appreciation both for the accomplishments of its first 30 years, and for its potential to shape the decades ahead. It is a rare privilege to join with them in our shared journey.

Romain Murenzi
TWAS executive director

About TWAS

The World Academy of Sciences (TWAS) works to advance innovation and sustainable prosperity in the developing world through research, education, policy and diplomacy. TWAS was founded in 1983 by a distinguished group of scientists under the leadership of Abdus Salam, the Pakistani physicist and Nobel Prize winner. Today, the Academy has some 1,150 elected Fellows from 90 countries; 15 of them are Nobel laureates. Throughout its history, the Academy's mission has focused on supporting and promoting excellence in research in the developing world and applying science and engineering to global challenges. TWAS receives core funding from the government of Italy. TWAS is a programme unit of the United Nations Educational, Scientific and Cultural Organization (UNESCO); its funds and personnel are administered by UNESCO. The Academy is based in Trieste, Italy.

About the editor

Daniel Schaffer was the public information officer for TWAS from 1997 until his retirement in 2012. From 1997 to 2007, he also served as the public information officer for the Abdus Salam International Centre for Theoretical Physics. He has written extensively on issues related to science, technology and urban development and is the author of *TWAS at 20: A History of the Third World Academy of Sciences* (World Scientific, 2005). He currently lives in Keizer, Oregon, USA, where he is a recipient of an Encore Fellowship with the Oregon Environmental Council.

About the Author

Index

Printed in the United States
By Bookmasters